George Henderson
Editor

Human Relations in

the Military

PROBLEMS AND PROGRAMS

Nelson-Hall nh Chicago

Library of Congress Cataloging in Publication Data
Main entry under title:

Human relations in the military problems and programs.

Bibliography: p.
Includes index.
CONTENTS: Leadership: Hebebrand, R. C. Human relations in practice. Patton, W. C. The leader as a counselor.—Race relations: Dansby, J. L. Race relations at base X. [etc.]
1. Sociology, Military—Addresses, essays, lectures.
2. Interpersonal relations—Addresses, essays, lectures.
I. Henderson, George, 1932-
U21.5.H8 301.5'93 75-15835
ISBN 0-88229-264-1

Second Printing 1976

Manufactured in the United States of America

Contents

Preface

On June 30, 1974, the United States Army completed its first full year without Congressional draft authority. Based on this limited period of observation, it appears that the all-volunteer Army — as well as the other branches of the military — will be able to maintain its authorized strength through imaginative and vigorous recruiting programs.

A strong military force is a requisite for America's domestic freedom and international tranquility. And, of course, this freedom and tranquility are dependent upon each branch of service operating viable "people" programs. Therefore, it is important that the armed forces create and maintain programs that improve leadership and minimize human relations problems, racism and sexism, for example.

Only a naive person would say that the armed forces are not improving in their efforts to minimize human relations problems. Even so, much remains to be done. The proliferation of military directives and human relations training programs clearly indicates an administrative awareness of the seriousness of lingering problems. However, there are still too many military leaders using their assigned missions as excuses for maltreating their subordinates. This is especially true of the officers and noncommissioned officers who cling tenaciously to the belief that the armed forces must not coddle their people.

It is sheer folly to discuss human relations problems in the military without relating them to civilian problems. Each subculture — military and civilian — does indeed have impact on the other. Thus, solutions found in one subculture may be adapted to reduce problems in the other. For this reason, I have elected to focus on problems *and* solutions pertaining to seven topics: leadership, race relations, women's equality, military justice, drug and alcohol abuse, health care, and civil service.

While written for students in military training programs, this book should be of value to all persons concerned with devising techniques and programs for improving human relations. Each year, there is an increasing convergence of military and civilian leadership styles. Of particular interest to the reader should be the case studies written by men and women actively involved in the process of social change. By sharing their anxieties, frustrations, and successes, the authors of these studies give us valuable insight into possible ways that each of us can bridge the gap between behavioral science theories and institutional practices. I gratefully acknowledge the importance of these case studies.

It is significant to remember that the military branches have led the nation in racial desegregation, and they are now in the *avant garde* of human relations training. But I repeat, much remains to be done.

part 1
Leadership

. . . if, beyond the straight way,
The captive elements and the ancient
Laws of the earth break loose
Like maddened horses. And then a desire to return
To chaos rises incessantly. There is much
To defend, and the faithful are much needed.

— *Friedrich Hölderlin*[1]

Throughout the ages, the art of human leadership has held a basic fascination for people in many fields. There has been endless speculation by historians, philosophers, management specialists, and human relations practitioners concerning the qualities or conditions that have endowed some persons with successful leadership capabilities and excluded such capabilities from countless others. Both those in authority and those under it strive increasingly to diagnose by merciless dissection the corpses of dead purposes and organizational goals that result when leadership is unproductive.

STUDIES OF LEADERSHIP

In the past half century, the study of leadership has assumed critical importance because rapid growth in size and complexity of organizational structures has necessitated greater versatility and sophistication on the part of those in leadership positions. There is a growing demand for expanded knowledge of the techniques and processes of leadership. Better methods have been persistently sought to identify, educate, and develop potential leaders for the insatiable maw of twentieth-century bureaucracy. Traditional concepts have been modified significantly by the combined results of research in the social sciences, education, and management. The resulting knowledge from this interdisciplinary approach has stimulated the development of fresh curricula in leadership, management, and executive development in schools serving the military, commerce, and education.

The exercise of leadership, though complex, is a most common and natural behavioral phenomenon. It occurs whenever one person influences the behavior of others for a purpose. Unfortunately, the exercise of leadership may harbor inherent weaknesses. Walt Whitman commented. "There is to me something profoundly affecting in large masses of men following the lead of those who do not believe in men." History is strewn with the rubble of people and nations who have fallen victim to those who assumed leadership roles and then acted from unbridled self-interest or from inferior motives. When the power that attends leadership is exercised in ethical blindness, irreparable harm may result. Irresponsible leadership is one of the greatest hazards men and women are heir to, for they must survive in a world where often, as

Rabindranath Tagore wrote, "power takes as ingratitude the writhing of its victims."[2]

Leadership is exercised by board chairmen, assembly-line foremen, commanding generals, and squad leaders. It is present even without the trappings or sanction of rank or position. It has been formally described as an interpersonal influence exercised in a situation, directed through the communication process, toward the attainment of a specified goal or goals.[3] Every concerted group effort involves effective leadership to pursue and achieve established objectives. Man is a social animal, and civilization has inched forward on the shoulders of persons in groups who inspired others to cooperate in moving the wheel of progress for a common purpose.

The universality of the leadership phenomenon, the diversity of the human personality, and group dynamics and the situational data available to describe the "leadership" phenomenon all make their contribution to the complexity of leadership studies. Increased urgency is added to these studies by the ravenous demands made by expanding organizations that are the foundation of our technological society. Larger and more complex structures involving more and more human and material resources are finding traditional methods of leader selection and training less and less adequate. Ever-increasing attention is being given to the critical nature of selection and preparation of those who will assume the responsibility for leading and directing others. The importance of this selection and preparation cannot be overestimated. It has been conjectured that 90 percent of the forward progress of civilization has been accomplished by the 10 percent of mankind who have pioneered in a diversity of leadership roles:

> *There is always room for a man of force, and he makes room for many. Society is a troop of thinkers, and the best heads among them take the best places. A feeble man can see the farms that are fenced and tilled, the houses that are built. The strong man sees the possible homes and farms. His eyes make estates, as fast as the sun breeds clouds.*[4]

Some of the fundamental perspectives and tools with which to probe the mysterious ingredients of leadership have been furnished by early advances in the behavioral sciences, the use of empirical methods, and the growth of Frederick W. Taylor's "scientific management" concept of the 1890s. More recently (during the middle and late 1950s), the work performed by such men as Bernard M. Bass,[5] Fred E. Fiedler,[6] Donald C. Pelz,[7] Ralph M. Stogdill,[8] and Robert Tannenbaum[9] has been incorporated into departments of management, in industry, on educational campuses, and in government agencies.

The military services, because of their urgent need for leaders, have continued to be instrumental in fostering research, study, and instruction in leadership. Ramón Lopez noted that as an agency, the military is unique in that "it can offer an almost free training period for individuals in such areas as management, leadership experience, and character development."[10] A massive accumulation of data on leadership has been compiled by the armed forces in the last fifteen years. Laboratory and statistical techniques have been revised to permit the analysis of interpersonal relationships in a more precise and empirical manner. Augmenting these efforts, the academic disciplines in the universities have placed their special insights and expertise at the disposal of military leaders. Recent studies reflect the general agreement that the complexity of human interrelationships resulting in effective leadership can best be explored by surveying: (1) the study of the individual as a leader; (2) the social psychology of groups; and (3) the situational factors affecting human interaction.

LEADERSHIP THEORIES

Napoleon I said, "All great captains have performed vast achievements by conforming with the rules of the art — by adjusting efforts to obstacles."[11] Many theories have been advanced to formulate, add to, or amplify concepts of leadership. Alfred Adler suggested that innate in every person is the wish to be the "upperdog"; no one wants to be the underdog. Nietzsche went even further and proposed that the "will to power" is the prime motive in life. Jacob Bronowski provided poetic support for Nietzche's theory:

> *At the heart of our violence, in act or in feeling, lies the wish to show ourselves men with a will. But the complexity of society makes the man lose heart. Nothing he does any longer seems a skill to be proud of in a world where someone else always hits the headlines. This is a plausible picture, in despair*

> *of which men cheerfully join any private army which will offer them the ambivalent identity of a uniform; the right to salute and be saluted.*[12]

Theories of leadership range over a wide spectrum — from Thomas Carlyle's theory that great men determine the course of history to Vilfredo Pareto's concept that situations permit certain types of persons to become leaders.

In the development of leadership concepts, one prominent study has classified leaders according to the means by which they assume their leadership. The classification categories derived from this theory are (1) the *emergent* leader; (2) the *charismatic* leader; and (3) the *elected* or *appointed* leader. This manner of classification places great stress upon the *process* by which leadership is assumed within a formal or informal group.

Joy P. Guilford concluded, "In almost any group of social animals one can find leaders and followers. Most leaders have to work to attain their positions and hold them. The struggle to reach the top among men and nations is a never-ending one."[13] In informal groups, a leader often emerges spontaneously from the group because he asserts himself and is accepted by others as being capable of directing whatever action is required to accomplish an objective. For example, a dozen or so boys may decide to play sandlot softball. One boy, perhaps because of his skill or experience, may initiate the choosing of sides and decide the ground rules that will govern the game. Initially, this leader may have to shout down or otherwise convince some of the opposition, but he emerges as the one who influences others to get the game started; he is accepted as the emergent leader. This is a normal procedure for establishing recognized leadership within unorganized, informal groups.

Similar to the emergent leader is the charismatic leader. He is possessed of some "mystical" personal power that causes his followers to bestow upon him the right of leadership. The degree of his technical competence to accomplish the task at hand, as compared with others in the group, is not the deciding factor. He is, rather, the one person to whom the group unhesitatingly looks for guidance. His leadership also has a quality of permanence that the emergent leader only rarely achieves. History provides many examples of such magnetic personalities and the sway they held over those who willingly, and sometimes even fanatically, followed them. Most religious leaders are possessed of this charismatic quality; however, this group extends beyond oracle, poet, and prophet. Those who watched the fanatic acquiescence of the German people when Hitler shouted, "Anyone who sees and paints a sky green and pastures blue ought to be sterilized!"[14] witnessed a charismatic roar of approval from the masses who followed him. The only difference in response to another charismatic leader, Abraham Lincoln, resided in the volume of the vocal response of his followers — not in the commitment in their eyes — when he said, "Fellow citizens, we cannot escape history. . . . No personal significance or insignificance can spare one or another of us. The fiery trial through which we pass will light us down, in honor or dishonor to the latest generation. . . ."[15] The crucial variable is whether the charismatic leader is dedicated to achieving humane or inhumane goals. The charismatic leader may be found in either a formal or an informal group and retains his or her leadership as the group becomes structured or formalized.

Within formally organized institutions or groups, the process of waiting for the emergence of a leader is too imprecise to leave to chance. In instances where more orderly processes of leadership designations are required, formal institutional leaders are appointed or elected. Elected leadership in democratic nations is determined by the expressed will of the people. Such elected officials generally see their duty as taking those actions that will represent the people who chose them. In the military, leaders are appointed by formally recognized superiors. It is evident that appointed leaders have a very difficult role. Frequently, the objectives of those in higher authority are incompatible with the personal motives of the appointed leader and thwart his self-realization as a person, as well as interfering with his fulfillment of his duties. Peculiar to the military are stringent penalties for disobedience. These present an added burden. A military leader must endeavor to achieve an acceptable compatibility between his unit's mission, which may be hazardous or distasteful, and the personal aims which he holds and the aims of individual members of his group. The speech of Kaiser Wilhelm II to his recruits typifies the dilemma of lesser leadership to higher leadership:

> *Recruits! Before the altar and the servant of God you have given me the oath of allegiance. . . . You have sworn fidelity*

> *to me, you are the children of my guard, you are my soldiers, you have surrendered yourself to me, body and soul. Only one enemy can exist for you — my enemy. With the present Socialist machinations, it may happen that I shall order you to shoot your own relatives, your brothers, or even your parents — which God forbid — and then you are bound in duty implicitly to obey my orders.*[16]

Every leader deals either with a group or individuals in a face-to-face relationship. The followers come to know their leader well. They form relatively valid opinions and attitudes concerning the leader's personality and capabilities and are directly influenced by his thoughts and actions. As an organizational hierarchy develops, however — as the leaders reach successively higher positions — the leader becomes less well-known to those who are affected by his decisions. His personality is perceived in an increasingly distorted manner and his influence becomes increasingly modified by the more immediate influence of the subordinate leaders. The possibilities of misunderstanding, modification, or error between an original order and its execution are abundantly clear and present one of the most common problems in effective leadership.

CHARACTERISTICS OF LEADERSHIP

Ralph Waldo Emerson wrote, "There are men who, by their sympathetic attractions, carry nations with them, and lead the activity of the human race."[17] It has been the tendency of humans to set their leaders apart as somehow different from ordinary persons. To probe the properties of leadership, it is only natural that prime consideration should be accorded the leader himself. This particular leader-focused approach to analysis of leadership has been called the "great man" concept. It is based on the theory that history can only be explained in terms of the great leaders who have effected changes in the history of mankind. Another prevalent concept has been that leadership is hereditary. The institution of royalty and the science of genetics are frequently offered and cited as proof of leadership being hereditary.

Other leader-oriented theories focus on the personality of the leader. Those who espouse these theories see the personality as composed of many different characteristics or traits. Individuals differ from each other to the degree that they display clusters of these traits. When attempts have been made by scientific research to isolate the traits of leadership, however, the results have been inconclusive in that few consistent trait patterns have been discovered. There is evidence that there is no single trait that consistently differentiates leaders from followers, although some dominant characteristics have been found. As an example, research data indicate that leaders tend to have higher intelligence than followers, but some evidence indicates that if a leader is too much more intelligent than his followers, his effectiveness as a leader is impaired: "All successful men have agreed on one thing — they were causationists. They believed that things went not by luck, but by law."[18]

Although invariable leadership patterns have not been established, traits and characteristics provide a means for generalizing about the leadership personality. Though it is true that there are no universal traits, certain traits are required by leaders of specific groups or in specific situations. The traits or characteristics required of a leader tend to fall into two general categories. Within the first category are those qualities required of those operating from a social and moral orientation; such qualities include integrity and maturity, which assist a leader in establishing the proper relationship and emotional climate with his followers. In the second category are those qualities supporting the capacity to deal effectively with the problem-solving and task organization confronting the leader's group, such as intelligence and judgment.

The United States Army *Field Manual 22-100* states that "leadership traits are distinguishing personality qualities which, if demonstrated in daily activities, help the commander to earn the respect, confidence, willing obedience, and loyal cooperation of his men." The manual lists bearing, decisiveness, dependability, endurance, enthusiasm, initiative, unselfishness, integrity, judgment, justice, knowledge, loyalty, tact, and physical and moral courage as those qualities to be sought in training men and women for effective leadership. Despite the most conscientious leadership, some persons, because of personality or environmental circumstances, descry even democratic leaders as despotic and make wholesale condemnations of all leaders. Oscar Wilde said, "There are three kinds of despots. There is the despot who tyrannizes over the body. There is the despot who tyrannizes over the soul. There is the despot who tyrannizes over the soul and the body alike. The first is called the prince. The second is called the Pope. The third is called the People."[19]

RELATIONSHIP BETWEEN LEADER AND FOLLOWER

Leadership cannot be studied apart from a consideration of those who follow and an understanding of the characteristics peculiar to the group being led. Attempts in recent years to move in this direction have been carried out through studies of social psychology and group dynamics. Both studies have resulted in rapid progress. While it has been recognized since the time of Aristotle that humans are essentially group-oriented, it is only recently that researchers have attempted to evaluate objectively and scientifically the interactions and dynamics that operate within groups.

Every leader participates in a number of significant ways as a member of the group or the contingent he or she leads. To achieve effective leadership, a leader must understand the significance of the group to its members, and the characteristics typical of group action. No person lives completely outside a group; both leader and follower contribute to, and absorb satisfaction from, the group of which each is a member. In the same way that the leader contributes something of his personality and characteristics to the composite personality of the group, the group assists the leader in achieving identity and in defining his or her role.

Members of military groups, because of the intense stress they must be able to withstand, require a high degree of solidarity, mutual confidence, and identification with other group members. Every military person, whether he or she is a leader or follower, not only is an individual within a group but also lives in a highly mobile situation which changes from time to time. As individuals, all GIs bring to these situations certain attitudes, frames of reference, and unique personality composites that are the result of past learning and experiences, which cause them to perceive their social and physical environment in a particular way. Group members, or followers, will have a specific response to a situation that is in accordance with the sum total of their unique characteristics as they interact with their group members in their environment.

In the process of interacting with the group, the follower ideally gains a sense of belonging. If proper leadership is exercised, individual followers come to identify closely with the group and attain a sense of solidarity and oneness. In return for these benefits, the group participation exacts individual compliance with group norms, fulfillment of formal and informal roles that the group prescribes, and contribution toward attainment of group goals. The bond between individual loyalty and group loyalty is a very tenuous one. Ariwara Ukihira cautioned, "The robe of mist worn by the spring — how thin the weft: by the mountain wind, so soon disordered!"[20]

As groups develop, both leaders and followers learn characteristic ways of handling varying situations. The interaction of group members determines the influence, communication, and friendship patterns. These patterns of interaction, individual member characteristics, and goal-orientation determine the direction and fulfillment of task performance. Analysis of groups over a long period of time has shown that informal group norms are factors that cannot be disregarded by the formal organization. Whether a leader can become effective in any group depends in part upon his or her clear understanding of the group and its members. This is not an easy task, for humans are not "fueled" by pure reason. Or as Francis Bacon observed, "Numberless in short are the ways, and sometimes imperceptible, in which the affections color and infect the understanding."[21] To gain understanding requires a considerable study of the personality of each group member (with his motives and past experiences), a consideration of the peculiar makeup of the group as a whole, and an analysis of the situation in which both exist.

An analysis of leadership must take into account factors external to the leader, for he cannot operate in a vacuum but mingles continually with his followers. It becomes evident that the complex social interaction between the leader and followers — and also the interaction among followers with each other — are vital factors. It has been theorized that it makes little difference what personality traits a leader possesses so long as the followers have faith in him.

In the military setting, a GI looks toward group members for the satisfaction of many of his personal needs, for in many cases he or she has been removed from the societal setting where these needs would ordinarily be satisfied. Albert Einstein said, "Everything the human race has done and thought is concerned with the satisfaction of deeply felt needs and the assuagement of pain. One has to keep this constantly in mind if one wishes, to understand. . . ."[22] In the leader-follower relationship, if the leader fails to satisfy the follower's needs, someone else will be sought for this purpose, for the drive to satisfy needs is one of the strongest human urges. If the follower must turn to someone else for need satisfaction, the institutional leader may be replaced in the eyes of the follower by an informal leader.

The success of a military leader in satisfying follower needs is hampered by the requirement that he or she orient the goals of the group so as to accomplish not only personal leadership goals, but also externally assigned missions. The consequences of failure are grave, for when leaders prove ineffective at either task, they can motivate followers only at a minimum level, and morale will be low, *esprit de corps* absent, and teamwork difficult.

LEADERSHIP IN RELATIONSHIP TO SITUATION

An often-heard expression in the military is, "It depends upon the situation." This expression implies that many situational factors external to the leader or his group influence his decisions and actions. Military leadership itself, compared to other forms of leadership, may be considered as particularly situational. Battlefield contingencies impose conditions in which standards must be flexible. An officer's oath of office and commission bear heavy moral responsibilities. Leaders are expected to set the highest standards of moral conduct, and in addition, they must take care of the personal problems of the men and women in their charge under trying situations. These responsibilities are far greater than those found in most commercial or civic leadership situations.

The military leadership picture is affected by an infinite number of variables, including mission, climate, geographical area, enemy, resistance, duration in combat, casualties, replacement, rest, state of training, and availability of equipment, to mention but a few. Some of these variables may be structured and controlled by sound leadership techniques. Each of three factors — the leader's traits, the group, and the situation — contributes to an understanding of the leadership process; no one of them is sufficient to explain the factors of leadership completely. Napoleon, for example, was able to control most of these factors: "Of the 60,000 men making up his army at Eylau, it seems that some 30,000 were thieves and burglars. The men whom, in peaceful communities, we hold if we can, with iron at their legs, in prisons, under the muskets of sentinels, this man dealt with, hand to hand, dragged them to their duty, and won his victories by their bayonets."[23] Clearly we do not expect our leaders to use these techniques.

INTEGRATED CONCEPT OF LEADERSHIP

The leadership trait theory is only a portion of the integrated concept of leadership, for it does not contribute directly toward a solution of problems. Whereas leadership traits and characteristics are relatively static, leadership is a dynamic activity. Traits and characteristics nonetheless have a strong influence on others, and they serve as excellent guidelines for the development of a leadership paradigm. Group dynamics theories are a requisite part of the integrated concept of leadership, for a leader always operates in a leader-follower relationship. A salient point in this connection — and one too often neglected — is the fact that the follower is not a mere automaton carrying out the leader's desires to the best of his ability. He is a human being with motives and goals of his own. In the rush toward the goal of achieving objectives, some leaders proceed as if it were proper that "things are in the saddle and ride mankind."[24] Such an attitude does not foster the strong attachment of followers to fellow group members and to their leader, which ultimately is a factor which will detract from goal accomplishment. The leader must recognize the existence of a multiplicity of individual and group factors that highly affect his ability to influence the group.

Leadership situations in the military differ markedly from those encountered in commercial or other service areas. The leadership styles appropriate for the many radically unique situations encountered by the military leader leave no room for the myopic or rigid personality. Each situation will influence a good leader's perspective and his techniques; he must be open-minded, adaptable, and responsive. The most logical method of analyzing the study of leadership process is an interrelational one that takes into account the interaction of all three basic factors: (1) the leader, (2) the group, and (3) the situation. Leadership necessarily is a dynamic interaction process involving the leader with his or her own personality, the group with its particular characteristics and needs, and the situation and its requirements in which the leader and group are operating.

LEADERSHIP STYLES

Kurt Lewin, Ronald Lippitt, and R.K. White conducted a classic study of leadership styles in autocratic, democratic, and laissez-faire climates.[25] The results of their study indicate that groups react more favorably to democratic leadership by displaying enjoyment in the task and by continuing to function effectively even in the leader's absence. Autocratically-led groups display more hostility, lower morale, and tend to fall apart during the leader's absence. However, quality and quantity of work done in the presence of autocratic leaders tend to be better than work completed with a democratic leader present. Under the laissez-faire leadership style, groups demonstrate little except boredom, horseplay, and either apathy or hostility.

Lester Coch and John R.P. French found that resistance to change is both individually induced and group-influenced.[26] They noted that frustration and anxiety were keys to *individual* resistance whereas social pressure, peer-group norms, group conformity, and cohesion were the forces of *group* resistance. Attempts were made to create a change by the method of group participation. By allowing the workers to participate in a decision concerning their work, change was effectively introduced, and productivity generally increased. Coch and French's study lent substantial support to those who encourage democratic leadership.

From the two studies, it appears that a democratic style is more desirable and effective than an autocratic one in certain situations, at least as far as subordinates are concerned. A democratic style is useful in guiding and coordinating a unit's thinking toward a group decision that insures a high degree of commitment to a mission. The group, in a sense, rules itself and the leader is accepted as a part of the group. Lao Tzu advised:

A leader is best
When people barely know he exists,
Not so good when people obey and acclaim him,
Worse when they despise him.
'Fail to honor people,
They fail to honor you';
But of a good leader, who talks little,
When his work is done, his aim fulfilled,
They will say, 'We did this ourselves.'[27]

Thus good leadership includes accomplishing an objective by making the unit's mission coincide with the goals and interests of its members. Fred Fielder[28] and Edwin Hollander[29] in their studies of leadership styles, however, found competence to be more important in determining a leader's influence than was either his style or his position.

Some of the foregoing findings pose a problem, for, regardless of how many individual personality differences exist among military commanders, the military remains an institution built upon executive authority. Unlike the society it serves, it is not democratic. Military commanders are not chosen by ballot. There are certain features of military operations that can never be democratic, primarily because missions most often are determined by a higher authority. The leader cannot always make the mission coincide with the goals of the group; therefore it remains for him to use his influence toward bringing the group to accept the unit's mission as one of their goals. The military is not likely to have the freedom Edward L. Benays described: "The freedom to persuade and suggest is the essence of the democratic process."[30]

LEADERSHIP CONSTRAINTS

A number of constraints act on the leader during the course of his job and inhibit him from doing his job with maximum freedom. One of the constraints is the military as an institution and its expectations of leader performance. What does the military expect of its leaders? Does it expect leaders to be hard and rigid? The important thing is how leaders perceive what their military branches expect of them. Evidence indicates that there is a goal conflict between superiors and subordinates. Most superiors want the leader to be task-oriented — to get the job done. Most subordinates want warmth, empathy, and consideration from their leaders. Basically, then, two terms are paramount in analyzing leadership styles: (1) *consideration* and (2) *initiating structure.* Consideration means taking care of the group under the leader's charge. Subordinates want from their leaders rewards for good performance, willingness to stand up for them, approachability, assistance in the solutions of personal problems, cooperation in keeping them informed, and behavior that is indicative of warm, trusting relationships. These are areas in which many young subordinates of today are saying that their expectations are not being met.

Initiating structure is another way of saying, "getting the job done." Specifically, it involves defining a mission, organizing the tasks to be accomplished, and devising methods to perform them. It includes such specifics as establishing organization patterns, developing channels of communication, and assigning specific tasks to individuals. One common error in selecting leadership goals is the belief that concern for mission achievement and concern for people are mutually contradictory. There is no yardstick of leadership extremes with consideration of individuals at one end and accomplishment of mission at the other end. The good leader will satisfy both.

One serious oversight has been the absence of processes that reinforce the concept of the interrelatedness of motivation and performance. Training techniques that facilitate arbitrating disputes, counseling subordinates, improving working conditions, rewarding performance, encouraging teamwork, and developing group cohesiveness are all-important for leadership development. The isolation often imposed by circumstances of military duty is reflected for many GIs in the poignant words of Thomas Wolfe: "The whole conviction of my life now rests upon the belief that loneliness, far from being a rare and curious phenomenon, peculiar to myself and to a few other solitary men, is the central and inevitable fact of human existence."[31] To deny or ignore this human condition is fatal to productive leadership.

If leaders are to be mediators between themselves and subordinates, there must be parameters that are recognized at all levels of the military structure and are translatable into general guidelines. Unrealistic principles or ideals will not suffice. A mediating framework should derive from conditions as they are perceived by the members of an organization. Some of the components for the mediating function are set forth in the following list, which may be used to formulate a framework for an effective leadership climate:

Structure: The feeling people have about the constraints of their work situation.

Responsibility: The feeling of being one's own boss, not having to "run upstairs" every time a decision must be made.

Risk: The degree to which people feel they can take risks in operating and improving their part of the overall mission.

Standards: The degree to which challenging goals are set for people. The emphasis people feel is being placed on doing a good job.

Reward: The degree to which people feel they are fairly rewarded for good work, rather than only being punished when something goes wrong.

Support: The perceived helpfulness of supervisors and contemporaries in accomplishing tasks.

Conflict: The feeling people have that supervisors want to hear different opinions to get problems out in the open where they can be dealt with.

Warmth: The feeling of general "good fellowship" that prevails in the atmosphere.

Identity: The degree to which the individual feels that he or she is a member of the group and belongs to the organization. Such feelings are expected to stimulate individuals to make sacrifices for the group or organization that otherwise would not be made.

CURRENT PROBLEMS AND APPROACHES

It should be clear by now that regardless of how much power or formal authority an organization "confers" on its leaders, their "usable" power and authority are granted by their followers. The now all-volunteer armed forces must cope with circumstances of individual dissatisfaction and organizational inequity. Furthermore, such problems must be resolved with imaginative and humane solutions.

If the institutions that purport to provide leadership and that are considered to be the foundation of our society are to continue, today's youth insist that they must be personal, honest, and include the will of the people. Large numbers of today's young people are reemphasizing the value of the person — individual dignity, honesty, integrity, and compassion. These appear to be excellent values from which to motivate a modern, voluntary armed force toward insight and involvement in leadership roles. To those who presently feel the sting of inequities between themselves and those who would lead them, "in heart's perspective the distance looms large."[32]

Many statistics are available that are eloquently translatable into some of the frustrations currently being expressed by minority groups and women regarding inequities in leadership opportunities. Despite efforts of the armed forces to insure "equal opportunity" principles, many persons are burdened with the handicaps of backgrounds of extreme poverty. Racism and sexism have also made them unable to capitalize on opportunities. To offset these handicaps, efforts are now being made to employ more personalized approaches that recognize and develop individual styles of leadership. A human relations approach to leadership is finding favor in the military and is being implemented in some military installations to winnow common goals from apparently contradictory points of view. Rap sessions, films and speakers with well-defined messages, immersion in culturally-alien situations, and role-playing are but a few of the many tools available to leaders endeavoring to create a viable organization that will benefit from the different tastes and outlooks of its membership.

The foregoing has been but a brief sampling of the history, theories, characteristics, relationships, styles, constraints, and problems of leadership. These comments simply point out that the need for effective leadership is more critical than ever before. Studies disclose that there is a new spirit among many to dare innovation, risk folly, and deviate from procedure to solve problems that have no guiding precedent. The rigid conformist and the technocratic midget are on the way out — as is the psychic distance between leader and follower. Leadership is not a spectator sport; the best and the worst of persons enter the arena. Albert Schweitzer offered this challenge to leaders:

> *Let those who hold the fates of people in their hands be careful to avoid everything which may worsen our situation and make it more perilous. Let them take to heart the*

marvelous words of the Apostle Paul: "As much as lies in you, be at peace with all men." They have meaning, not only for individuals, but also for nations. In their endeavors to preserve peace among themselves may nations go to the uttermost limits of possibility, so that the human spirit may have time to develop, to grow strong, and to act![33]

Human Relations in Practice

Captain Roger C. Hebebrand
Third Mobil Communications Group,
Air Force Communications Group
Tinker Air Force Base, Oklahoma

I went through the Tinker Air Force Base inprocessing and acquainted myself with the people and my responsibilities, and I had conferences with my bosses, and so on. Looking back now, I find it extremely difficult to express my initial reactions — shock, anxiety, disbelief — for after what seemed normal enough inprocessing, I found myself a leader of a branch that had been and was operating in a fashion that was the antithesis of everything I had learned about leadership, management, and problem-solving.

DEFINING THE PROBLEM

I spent the first two weeks analyzing the situation. First, the branch had a highly centralized power elite. Leadership patterns were nearer the dictatorial end of the spectrum than the democratic one. The officer whose place I took had delegated nothing. He had made virtually all decisions. Communications had always been downward, never upward. Personnel, even those with the most experience, had been treated as pawns. Very little personal consideration had been given. Violators of the decrees that had been given were severely disciplined. I observed that morale was very low, and, of course, disciplinary actions were on the increase. There was a high degree of tension and conflict in all areas of my responsibility. My predecessor, I found, had very little to do with his troops. I had met him only briefly two weeks before for about an hour. I had made observations from this brief encounter: first, my predecessor, although businesslike, was distant, almost as if he were forcing himself to go through the ritual of seeing me. Second, when he took me around to introduce me to some of the troops, I noted a guarded look on their faces. Being new, I thought that it might just be the fact that I was going to be their new boss and that they were going to wait and see. Judgments would be made later. It was not long afterward that I was to learn that elements of openness and trust were not there.

Perhaps the most challenging person I encountered in my entire military career was my second-in-command, a first lieutenant. In him I found both success and failure. In temperament, background, attitudes, approaches to leadership, methodology in approaching and handling of conflicts and problems, he was the very opposite of everything I stood for. This one individual had managed to cause more negative behavioral patterns and disrupt normal flow of communications more effectively than five people would in their entire careers. He had managed to elevate problems to the general level that any aggressive, sharp staff sergeant could have solved. He had managed to create such a hostile environment that the effects had been felt at the Air Force Logistics Command (AFLC) and Air Force Command System (AFCS) levels. However, it must be pointed out that as in chemistry, certain solutions in themselves are ineffective for explosives, but mix them together with the right variables and a very powerful explosive results. In this state, extreme caution must be taken. Proper handling is a must; the slightest mishandling could be fatal. My research revealed that, yes, my second-in-command was primarily responsible for the condition in my branch. But for him to have had such far-reaching effects, other factors or variables had to be present. What were those factors? What kind of person was he? Why did he act as he did? Were conditions as bad "in house" as they were on the "other side" (the civilian employees on the base)? I needed this information if I was to develop a plan for positive change. I had to gather data, analyze those data, and create and execute a plan with careful precision if I was to be successful. Then I could see whether human relations skills and theories were practical in real-life situations.

I spent two weeks gathering as much information as possible about my second-in-command from observation, interviews, and other sources, such as records and files. Finally enough data had been gathered to help me understand the whys of the lieutenant's behavior. This, his first assignment had unfortunately not been compatible with the system at most Air Force bases. It was very frustrating for him from the very beginning, and, moreover, flexiblity was not his strongest trait. Given other circumstances — the right boss, less responsibility, and more time for him to mature — things might have turned out differently. But, unfortunately, certain earlier events, the leadership patterns of the commander, a boss who was not really interested in the Air Force as a career, and was more interested in preparing himself for law school, were the conditions as I found them.

What were these earlier events? Shortly before the lieutenant's arrival, the previous commander had been fired. His unit had failed the Operational Readiness Inspection (ORI), the supreme test for any military organization to meet its possible mission contingencies and test its combat readiness posture. Whatever the reason — whether the commander had been incompetent or the victim of a weak staff — he had been relieved. A new commander had taken his place. Very dynamic, extremely demanding, he tolerated no deviations from his command. He was not going to be fired. He would tackle any problem head on and would fight anyone who had not, or seemingly would not, support his mission. This included that huge system, AFLC. He did not have time for diplomacy, nor could he afford to consider individual feelings; he had a mission to support, and failure he would not tolerate. I do not mean to be critical of the commander's approach. He did by sheer force of personality accomplish the command objectives and performed certain feats of unparalleled excellence. Upon his retirement, he was awarded the First Oak Leaf Cluster to the Legion of Merit, the award second only to the Congressional Medal of Honor.

Owing to the fact that the officer in charge, only recently promoted to captain, was occupied with his educational pursuits, most of the day-to-day activities, personnel activities, coordination, and system implementation were left in the hands of the lieutenant. The lieutenant tried desperately to convince his superiors that he should be allowed to take over when the captain was discharged from the Air Force. But the captain had convinced the commander that the job demanded a higher ranking officer with more experience than he or the lieutenant possessed, and the commander went along with the idea. The post was upgraded from that for a captain to one for a major, and the commander requested a senior captain or major from AFCS to fill the vacancy.

The lieutenant felt that he had been betrayed. In his view, he had given everything to earn the right to be in charge. When I arrived, resentment, although never surfacing consciously, was characteristic of his actions. From the very beginning, it was apparent that he was going to fight me, though not openly, for his obsession with rank and authority never permitted the openness that I wanted so much. I also represented an attitude that he had never before experienced. My approach to leadership was so different that he just could not accept it. His mind was closed.

In any system as huge and as complicated as the Department of Defense (DOD), logistics system problems are bound to occur. The degree to which these problems are solved depends to a large extent on the skill, initiative, and motivation of the managers involved. However, no matter how self-evident this may be, the most difficult task for anyone at any level of management is defining what his problem is. To the inexperienced this seems obvious, but to the seasoned veteran, it is far more difficult than it first appears.

Several times during those first few weeks I asked the lieutenant to define the individual problems, as concisely and as clearly as he could, in several different areas in which we were experiencing difficulty. If I heard it once, I heard it a hundred times: "Those damned civilians [nonmilitary employees at the base], Captain, are the problem. I don't trust them. They are different." He always referred to "them" as the problem. In other circumstances, it would have been funny. I simply could not communicate with him. It was much like an Englishman trying to talk to a Russian and neither man understanding the other's language. I am sure that it was as frustrating for him as it was for me. The system is mechanically somewhat different on an Air Material Area (AMA) than on the regular Air Force base. Most of the Air Force utilizes what is known in supply circles as the Standard UNIVAC-1050-II computerized supply system. The AMA is based on a DO-33 computerized supply system. Although overall supply knowledge and language may be sufficient — or at least should be to have a vehicle for common communication — more often than not, it is not. Common sense and flexibility are two indispensable realities of the manager who gets things done. I began to see some of the whys of the clashes between the AMA and us, for my second-in-command lacked both qualities and had made no attempt to understand the system he was depending on to support him.

The vehicle the lieutenant had used to "force" support was a letter he wrote for the commander to sign. The commander in turn sent the correspondence to the general at AFCS. The general would send his response to the general at AFLC, who would send it down to the general commanding the AMA, who would request an answer from the applicable directorate, and so forth. The smallest problem was blown up beyond all proportion. This procedure created tension, conflict, and finger-pointing. It was almost unbelievable to me that anything got accomplished under these conditions. Even the basic day-to-day operating procedures within Material Control were clogged with bottlenecks. Petty office procedures were preventing effective accomplishment of routine operations. I observed that on many occasions the lieutenant failed to heed the advice of the more mature, more technically qualified noncommissioned officers within the branch.

INITIATING CHANGE

The first set of circumstances I changed was the decision-making process that had traditionally been held by one or two persons in charge. In doing so, I was making certain theoretical assumptions. In the vernacular of Abraham H. Maslow, I assumed that most persons prefer to be prime movers rather than passive helpers, tools, corks tossed about on the waves. I assumed that persons would perform better if they believed that they had a part in the way things happened. I assumed that each person assigned to his particular function would develop a sense of worth and a sense of his uniqueness in having the opportunity to make choices in the decision-making process, hopefully resulting in high production and morale. However, caution would have to be rendered at this point, for too much responsibility can crush a person, just as too little can cause boredom and lessening of initiative. I had to believe that people would rather build and create than destroy and tear down.

Even with these basic assumptions, which I believed were essential, I simply did not have the time or the energy available to have all decision-making placed in the hands of a few — mostly me. Besides, I certainly had no delusions or fantasies about my indestructibility or an exaggerated sense of self-importance. I merely had a responsibility and a job to do. However, I by no means lacked a sense of ultimate responsibility or of the importance of my position. Nor did I lose sight of the ultimate objectives I was working toward. More than once I had to modify some of my basic assumptions. When I made this change, I was extremely open. I gave my staff members my overall philosophy of leadership, my ideas of what I believed our ultimate goals were, and a rare gift of personal choice.

From the start I made it clear that this new approach was not to be interpreted as meaning there was a weakness in my style of leadership. I told my subordinates frankly that I was capable of dictatorial leadership and could effectively execute that power if I was forced to.

To some human relators, coercive leadership patterns and limitations of any magnitude in personal decision-making are apparent contradictions to basic principles of human relations. But because my approach was so different from what most of them were used to — particularly in the military — and because certain rigid behavioral patterns had been preset and conditioned by past leadership techniques, certain limits had to be understood and limited coercive measures taken so that the thawing of these negative patterns would not be too drastic and at the same time would be directed toward positive ends.

To this basic move, the first lieutenant voiced his opposition strongly, advising me that leaving choices to my middle managers and lower-ranking airmen was dangerous, that it would not work, that the commander's wrath would fall on me when I failed, and that most certainly I would fail. He gave me two weeks. I smiled. I told the lieutenant that I was willing to take that risk and that the responsibility for the success or failure of my activity was mine alone. I told him that I appreciated his concern but that I wanted him to follow my policy. Incidentally, all that two-week time frame was an interesting one; the lieutenant was always giving me this time limit to show me how long my ideas would work in the organization. I am not sure of its significance even to this date.

I sensed an optimistic outlook on the part of my other staff members, however. I treated them like human beings. I never put up my rank to them at any time. and I would talk as casually to the lowest-ranking airman as with the highest-ranking NCO and imply an unspoken trust and confidence in each one. This helped break down the guarded attitudes, and my people for the first time felt that they could express their feelings without fear of disciplinary threats.

STABILIZING THE CHANGE

It is necessary to explain here that the first step in my plan as a change agent called for reforms within my own branch before I could effectively work with the giant just waiting outside our gate, Oklahoma City Air Materiel Area (OCAMA). At this point, I should interject another important note. I wish that I could say that I found "the structure" that would prove at all times to be the best. But if there is one lesson I have learned in my experiences as a human relator, it is that very few ideas, if any, are absolutes. They are relative. A more realistic conclusion born out of my experiences would be "situational." My overall ideas have remained relatively stable, but I have modified more than once my original steps to change.

I had read rather extensively in the area of organizational planning. I had found out that, although there were many conflicting ideas in this area in the literature, there were certain threads of agreement that were helpful to me in making my decisions. I realized that I still needed to be cautious in how I employed the generalities expressed in what I read as principles. I had found many useful lists of organizational-planning principles, but perhaps the simplest and shortest series was that of James L. McCamy in *Science and Public Administrative*. These basic rules say a great deal in just a few words:

1. Definite, clear-cut responsibilities should be assigned to each executive (person).
2. Couple responsibility with authority.
3. No person occupying a single position should be subject to orders from more than one source.
4. Give the section heads all the staff services they need.
5. No administrator should have reporting to him more subordinates than he can supervise adequately.

6. The main subdivisions of organization should be based upon analysis of activities, and activities that are alike should be put together.
together.

Along with this list were several other ideas. How much decentralization was really practical? If I decentralized too much, authority in decision-making might go too far afield from my general policies and might call for more authority. I found the words of Henri Fayol, a well-known management writer, most helpful: "The question of centralization or decentralization is a simple question of proportion; it is a matter of finding the optimum degree for the particular concern." The span of control was critical. The number of people reporting to me should not exceed the number that I could direct and coordinate. I wanted everyone to know who his boss was and who his subordinates were, but I also wanted each person to feel that he was important to me. How was I to insure that all the people up and down the line would effectively communicate with each other? As a service group, we had to respond to requests from many departments, and oftentimes divisions of responsibilities were not clear-cut.

But I believed that there are at least three general attributes that characterize any good organization: *balance, stability,* and *flexibility.* An organization in balance is one whose components have been allowed to develop fully. Of course, organizational objectives must be considered in this development. Overgrowth or stunted development should not be permitted, and no element should be granted more or less emphasis than is proportional to its importance and value. Stability in an organization indicates that it can develop and sustain itself without overdependence on individual key staff members or other specialized personnel. No man is indispensable, and for all positions there is a way of filling in losses or absences. All levels of personnel should be prepared to step up or stand in as the situation demands. Many factors contribute to flexibility, but essentially it is that quality which enables changing workloads or unforeseen problems to be handled quickly and efficiently with a minimal amount of conflict and frustration.

With all this in mind, I explained my ideas in management, how I expected personnel to be treated and the goals I was striving for. My greatest obstacles were the lieutenant, as I have indicated, and a technical sergeant who was in charge of a section that was very important to the mission of our unit but who, although having spent many years in the service, did not agree with me, lacked insight into what I was trying to do, and had certain unfortunate personality traits. These obstacles were particularly difficult when we were handling delicate situations with OCAMA.

This brings me to my second area of change — effective structuring of my organization.

The situation I had inherited presented a most interesting challenge. My staff was composed of two officers, a second lieutenant and the first lieutenant. At first, it appeared that the line structure was well-defined on paper. The chain of command had at the top the branch chief, or the chief of materiel (me). Under me were two major areas of responsibility — supplies and equipment. This is an oversimplification, but it will suffice for this discussion. Structurally, on paper, the first lieutenant was a subbranch chief, responsible for the supply areas; the second lieutenant was responsible for equipment. Under each subbranch were a number of smaller units headed by various levels of rank. In actuality, however, Materiel Control did not operate as it was depicted on the chart, and areas of responsibility were in many instances duplicated, ill-defined, overlapping, and confusing. If the reader of this paper is confused at this point, imagine this one human relator trying to make order out of this chaos.

The second lieutenant, I found out, was really a mere figurehead. Impossible as it might at first appear, he had been assigned to the equipment subdivision for six months before the senior ranking NCO in Materiel Control found out that the second lieutenant, not he, was in charge of the equipment area. The commander had used the second lieutenant in several special assignemnts. He was in reality an "overage." The second lieutenant was much more levelheaded than the first lieutenant, more down to earth, responded less emotionally in most areas to problems, and had a higher tolerance level to frustration. However, he had very little understanding and awareness of the real situation in Materiel Control. This was not unusual, since he had had virtually no experience in supply and since the commander had kept him occupied with other assignemnts. Although it had indeed been an honor for the commander to use him on some important projects, he had not been exposed to his actual career field. He had only been on the job for two months. The senior NCO, only one grade below the highest-ranking enlisted grade possible, ran the section. The second lieutenant, for all practical purposes, might as well not have existed. Further analysis proved that even this highly skilled, extremely competent NCO had his hands tied and was virtually powerless. His counsel was not heard nor his recommendations followed. He had no authority. He had had twenty-four years' experience, was a senior master sergeant, and was being used as a high-class clerk.

The time had come when diplomacy, counseling, and communicative skills had been thwarted. For a brief time I, the human relator, became a dictator and through mere force of authority vested in the position *ordered* the following:

1. Under no circumstances would the first lieutenant be allowed to coordinate with any member inside or outside the group concerning supply or equipment matters and
2. until further notice, the first lieutenant was relieved of virtually all power of authority. He would perform only those projects I selected for him.

I also went to the commander and asked that the first lieutenant be transferred to another Air Force unit, since his effectiveness had virtually been destroyed and was irretrievable even if he did a 100 percent reversal of his past performance. I told the lieutenant why I was going to take this step long before I did so. The first lieutenant was gone within thirty days after my conference with the commander. I tried to the end to help him see himself as others saw him, but I failed. Long after I am forgotten by him, I will wonder whether he could have been reached and, if so, by what means.

The next logical step, if I were to have followed the line structure to the letter, would have been to make the second lieutenant next in command. This I did not do. Instead I created a temporary position and placed him in a position lateral to me as my special staff. In essence he was to be my understudy until I thought that he was experienced enough to take over in my absence. I also selected projects designed to familiarize him with the basic concepts of diplomacy, problem-solving methodology, and supply procedures. Furthermore I wanted him to become familiar with the DO-33 computerized system of the AMA, its advantages and disadvantages.

The key move in my plan came when I named my senior ranking NCO as the overall superintendent of the Materiel Control Branch. To me he was the very epitome of competence, loyalty, and devotion to duty. No officer could have asked for more. He did not always agree with me, particularly in some of my techniques of human relations. He had my complete backing to disagree at any time. I truly respected his judgments. In actuality, we complemented each other. In only rare cases did I not take his counsel, and in those cases he carried out my policies to the letter.

Shortly after the first lieutenant left, I gave the technical sergeant, described earlier, a choice. He could retire or be fired. He also refused my counseling, and in addition he had personal problems beyond my area of competence to handle. I replaced him with a newly assigned master sergeant. Other changes were made. We tried to match the right individual with the right job; to match maturity, experience, and capability; to minimize failure and frustration; and to promote a sense of dignity in each staff member. Although a multitude of factors were considered, I tried to act on two basic assumptions. First, each person must feel that his job is gratifying, that the content of his job is both interesting and worthwhile. This factor many times is determined by individual supervisors and how well they can communicate. Second, there is a direct relationship between intellectual and personal freedom in job productivity. In nearly all instances, an earnest effort was made to see that each person was granted those freedoms.

The next step in my plan centered on certain internal practices, both methodological and philosophical, that I believed were preventing optimum results.

I had noticed that the first lieutenant had spent an unbelievable amount of time collecting information, writing correspondence of a negative nature, and, finally, spending much of his time anticipating what the commander might want. This activity, of course, involved not only a great deal of his own time but of that of the senior NCOs in the branch. After thoroughly investigating these facts (doing a kind of time-and-motion study), I set down some firm guidelines. Not one letter of a negative nature would go forth from the branch until further notice. All problems would be approached in a logical manner: the problems would be defined, data would be collected, and solutions would be recommended and jointly resolved, whether with OCAMA or with our own organization. There would be no more time spent anticipating what the commander *might* want. It was my conviction that if he wanted some information or some task performed he would say so.

That was the easy part. Changing attitudes and general philosophy was not so easy. I experienced both success and failure. It would probably be realistic to say that I modified attitudes enough so that at least my staff members were not pointing fingers at "them" across the base (OCAMA). I tried to keep my staff members from looking for scapegoats and to help them realize that there were problems on both sides and that the only way they were going to be solved was to emphasize a common base and approach problems jointly. I pointed out several instances in which we ourselves were creating many of our problems. Further, if any criticism of OCAMA was to be made, I, not they, would make it.

The next phase centered on the personnel. Each person was to be treated fairly. Supervisors would first ask, not order, people to do things. I wanted an atmosphere of trust, the absence of hostility, and a minimum of frustration. Before disciplinary measures were taken, evidence had to be provided that maximum effort had been made to help the individual airman understand what was expected of him and the standards to which he was supposed to conform. Nothing was to be assumed. I wanted to interview personally every new man and every older one. Each supervisor was made responsible for seeing to it that each man under his supervision would be made thoroughly aware of the importance of his particular job. Finally, I wanted honest feelings of cooperation to be generated at all times, both to our customers in our own group and in our relations with our civilian counterparts in OCAMA.

BRANCHING OUT

Now I considered the next step. My attention turned to the AMA itself. While the basic changes I made in the aforementioned steps were occurring, other events were taking place at the same time.

I made an appointment to see every major section and branch chief in the areas on which we depended most for our support. I told them who I was and where I came from. I do not believe that I ever experienced such guarded looks or general hostility as I did with each one on our first encounter. One of the most effective tools that I have learned through past experience in a situation like this is humor. It eases tension and clears the air. The next step is to ask each person for his account of things, letting the chips fall where they may. I listened very closely to the men, trying to stay away from anything that might trigger a defensive attitude.

I gave each chief my philosophy of management. I assured him that the way things were handled in the past was over. No negative letters would be sent. All problems would be approached jointly for a solution. If any problem came up in which a serious disagreement occurred, we could send it to the next higher level with joint effort and coordination. Somehow the conversation always managed to come around to the first lieutenant who had not yet been transferred. I assured each chief that he need not worry about the lieutenant. I would be responsible for all major coordination at first. After explaining my philosophy of management, my approach to problems, and the methods I was following in my own unit, I asked each of them a favor: if they were favorably impressed with my approach, tell my commander either verbally or in written form. This was not to be done for my personal advantage careerwise, but it would put the commander in a positive frame of mind and, as such, strengthen my approach. It would also plant seeds of mutual cooperation and understanding. It would be my task to report positive ideas and, most important, results to the commander. Where the lieutenant had been negative, I had to plant positive seeds. I emphasized commonalities rather than differences.

Naturally, it was going to take more than any mere first interview to undo all the hostility and negativism. Although they all seemed to like my approach, I sensed a "wait-and-see" attitude.

I spent a lot of time educating myself to the way things were done under the DO-33 computerized system, something those in charge previously had never attempted to do. If I had to depend on the system, and if I was to be successful, this was a must. During the weeks that followed, it proved a most rewarding effort and proved very effective for cultivating a basis for communication. Each person I came in contact with agreed that change was welcome. My approach with each major staff group was the same: (1) friendliness, (2) humor, (3) openness, (4) discussion of common objectives, (5) a summary of my personal approach to and philosophy of management, (6) commitments from each staff about the reasonableness of my request, (7) an honest attempt on my part to understand the other staffs' problems and limitations, (8) an agreement on a mutual approach to problem solving — face-to-face, personal interaction and no impersonal letters, and (9) recognition of those who put forth extra effort. The plan worked.

AFTERTHOUGHTS

Though this sounds like a success story, there were and are regressions to old ways. Oh, yes, those who "count" see the change and are amazed. But one of the important questions about any process of change is whether or not the change will remain a stable and permanent characteristic of the system. Too often change that has been produced by painstaking and costly efforts tends to disappear after the change effort ceases, and the system slips back into its old ways. Retraining of old thinking habits, constant reviewing of problem-solving techniques, the importance of communication, and the reminder of what research has proved in the areas of human relations are a continuing process — the job of the change agent, the human relator, is never done.

I count myself fortunate indeed to have gone through this experience. Few students of human relations find themselves so early in their careers exposed to such real-life laboratory conditions where they can test their ideas.

This experience has proved to me beyond a doubt that human relations skills and techniques are by no means a mere intellectual answer to life's problems. Skillfully used, they can result in unbelievable rewards.

There are, finally, no absolutes in dealing with human beings, only general guidelines, each situation to be dealt with individually. The risk is great, but the rewards a delight. The change agent's job is not completed. Perhaps he is more skillful and better aware of the challenges in respect to the human being, nature's most challenging creature.

The Leader as a Counselor

**Colonel William C. Patton,
Assistant Chief of Staff
Marine Corps Recruit Depot, California**

CASE 1

As the Marine Corps Representative at the United States Army Field Artillery School (USAFAS), Ft. Sill, Oklahoma, I was accountable for the satisfactory performance or lack of success of all Marine Corps officer students (approximately 300 lieutenants per year) attending the 12-week, 52 hours per week USAFAS course. Since the students learn skills concerned with firing artillery, poorly learned lessons can actually be fatal to the troops they may be supporting. At times I was faced with the need to devise ways to motivate the students toward doing a better job of learning.

My predecessor had motivated all basic students through the use of *fear,* and few basic students ever wanted to go to his office for a conference when they were having difficulties with their studies. I attempted to change this image as much as possible, and my method appears to have succeeded — the rate of failure decreased more than 50 percent during my command.

When a new group of students arrived, I met them informally during their first week. I briefed them on policy, told them a little about the artillery course, and offered to answer questions concerning any problems they may be having. My primary objective in this session was to establish open, two-way communication channels between us. This session lasted as long as necessary but usually between 1-2 hours. In my closing remarks, I assured them of my interest in them and my concern for assisting them in any way possible in order to contribute toward a most meaningful educational period.

As the school program progressed, I monitored the Marine students' progress very closely. If an officer started to falter, I scheduled a conference period with him. During the conference, we discussed his problems and tried to devise solutions. In some instances, this was enough to assist a student in solving his academic problems.

Once a student reached a point in his graded work where he could no longer reasonably be expected to achieve the required passing score of 70, he was then considered for transfer from the school or for recycling into another class — which would permit him to retake portions of the course that he had failed.

I found that I achieved a higher percentage of administrative success with a recycled student if he requested to be recycled rather than my merely informing him that I intended to recycle him. Using the approach of presenting recycling as an alternative and then letting the student request it, I had only one recycled officer fail to complete the course satisfactorily. I feel that by placing the responsibility of a person's career in his own hands and supporting his selected option, I was able to contribute to a Marine's maturity as well as his motivation to succeed.

There is, of course, a material as well as a human savings in raising the course completion rate. It requires a significant financial expenditure to send an officer to the Artillery School. Both time and money are lost if a student fails to qualify.

I met again informally with the members of each class at the end of their course. The object of this meeting was twofold: first, I tried to get as much feedback as possible on the course and problems encountered and, second, I tried to answer all questions I could about each student's next assignment, usually his first real encounter in the regular forces. This session — also about two hours long — proved to be quite beneficial. An unexpected benefit was that many course improvements grew out of the constructive criticisms of the graduates. Because of the open channels of communication, most lieutenants stated that they had a chance to ask questions that they normally would not have felt free to ask their commander. I am convinced that when our graduates reported to their next duty assignments they did so with the confidence that they were able to perform well as an artillery officer. I would also like to believe that I modeled behavior that will guide them in being humane leaders.

ANALYSIS OF CASE 1

Fear can be a very powerful stimulus; however, I feel that the use of fear as the means to achieve superior performance did not contribute toward the maturation of the individuals attending USAFAS. Also, excessive pressure can cause mental blocking, which eventually will physically immobilize a person. Therefore my leadership style was designed to reduce the pressure on the student.

Believing that an open channel of communications with the students was very important, I tried to establish it as soon as possible. On many occasions, the students invited me to their informal social events. This offered even better opportunities for me to get acquainted with them as individuals. On all occasions of this sort, I tried to be a good model for them to emulate. When I exhibited enthusiasm for the school and what it taught, the students generally exhibited similar enthusiasm. This is an example of teaching values by living them.

When an officer began failing, I immediately conferred with his instructors in an effort to determine his attitude and whether or not he appeared to be trying to succeed. After I had this information, I scheduled a conference with him. During the conference, I was very careful not to intentionally make statements which would discourage the student or cause additional stress. The central focus of my conversation was upon the course in which he was having trouble, and possible sources of assistance. If I was told that the student appeared not to be trying to succeed, we discussed the accuracy of his instructors' perceptions. This conference, in most instances, was enough to get the young officer back on track.

When an officer was not able to keep his grades high enough to pass and reached a point beyond which there was little chance for him to recover, a second conference was held. During this period, we discussed the alternatives still open to him. At this time, I would talk in detail about his academic problem and the effect this could have on his career in the Marine Corps.

If necessary, I formally met with the student a third time. During this conference, we discussed the advantages of recycling him into another class that would allow him to repeat the material in which he was deficient. In most cases, the student in this condition would request to be recycled. If he did not request recycling at this time, I then informed him that he was going to be recycled and would have another chance. At all times I attempted to phrase my communications in such a way so as to encourage him and build confidence. The finished product, I believe, *must* be a confident officer.

Accountability is not new to the military. Because of the large expenditure of funds and time to send an officer to school, every attempt must be made to make sure that the expenditure is not wasted. My mission was to make the learning situation as conducive as possible to the students' becoming qualified artillery officers. Therefore, at times I was faced with the necessity of recycling students who did not request it. When this happened, I did not facilitate the most desirable learning situation.

In the final analysis, perhaps my approach is nothing more than a sophisticated authoritarian approach that results in a gradual buildup of fear — the fear of failure in the Marine Corps. Most of us want to succeed. The question is: At what cost?

CASE 2

Two weeks before Gunnery Sergeant (GySgt) E was due to report to Ft. Sill to become a member of my staff of instructors, I received a phone call from his last duty station. The officer on the line informed me that GySgt E had been involved with the wife of Staff Sergeant (SSgt) F while SSgt F was deployed to Europe for six months. The officer stated that SSgt F's wife had given GySgt E many items of expensive household furniture (color TV, home entertainment console, king size bed, and other smaller items). GySgt E, in turn, had this furniture packed and shipped to Ft. Sill as household effects and had supposedly run off with SSgt F's wife. The officer requested that I investigate the charge as soon as possible.

Approximately five days prior to the arrival of GySgt E, I received a personal letter from SSgt F restating all the above charges and also that he had hired a private detective and a lawyer to assist him in recovering his household effects. He also asked that I assist him as much as possible.

Because duty at Ft. Sill requires a clean record and excellent performance in a specific skill, I was tempted to pick up the phone and call Marine Corps Headquarters in Washington, D.C., and request that GySgt E's orders to Ft. Sill be cancelled and that he be retained at his present duty station until this matter could be resolved. This would have undoubtedly cost GySgt E the opportunity to serve at Ft. Sill as an instructor, which is considered a rare opportunity and career enhancing.

I decided not to call Headquarters and not to judge GySgt E until he had the chance to give me his side of the story and defend himself against these rather serious charges. I instructed my administrative chief to inform GySgt E that I desired to see him as soon as he reported to this station. GySgt E reported to me on April 10th.

As required by the Uniform Code of Military Justice, I informed GySgt E of his rights to counsel, his right to stand mute, and that any information he voluntarily gave could be used against him in a court martial. I then informed him of the charges leveled against him by SSgt F and I let him read SSgt F's personal letter to me. I explained how the whole thing appeared to me, looking only at one side, and that I would like to have his side of the story before I made a decision in this matter.

He said that he had nothing to worry about and that he would like to acquaint me with some additional facts that I apparently did not have. First, he said that he had been friendly with SSgt F's wife but had never cohabited with her as was charged. He said that actually he had been a friend of Mrs. F and her family, including her mother, for many years. He further stated that at no time had he lived with Mrs. F nor did he have any intention of doing so. He stated that Mrs. F was in the process of getting a divorce from SSgt F. She was selling all her household possessions and she had sold items to other people, too. GySgt E produced a receipt for cash he had paid Mrs. F. The receipt was for $450.00, which appeared to be a fair price for the furniture in question.

GySgt E stated that when he heard about SSgt F's anger concerning the sale of the furniture, he reported the matter to the local area police and asked their opinion as to the legality of the sale. He stated that he was told by the police that he was not breaking any laws. He said that he had also consulted with the base legal officer as to what steps, if any, he should take and was told to ship the furniture and forget the threats. GySgt E further stated that when SSgt F arrived home from his deployment, he attempted to contact him and explain the matter, but SSgt F had refused to discuss it with him.

I suggested that GySgt E consult with the local base legal office and acquaint them with the elements of the situation just in case it was necessary to ask for assistance from them in the future. GySgt E did as I had suggested. He further tried to contact SSgt F and talk to him. He was able to talk to SSgt F, who stated on the phone that he was actually doing a lot of things concerning his wife in order to make it difficult for her to obtain a divorce and if she did get the divorce it would be granted without a support requirement from him.

My opinion of the whole affair changed considerably and I am confident that my decision not to cancel GySgt E's orders to the school was a correct and just decision. Had I taken any other course, I would have caused a blemish on an otherwise fine record of an apparently dedicated noncommissioned officer.

ANALYSIS OF CASE 2

When a man's career is at stake, it is mandatory that he be given every opportunity to defend himself and his reputation. A leader should not make a decision without all the available facts. First impressions based on insufficient facts can result in grossly unfair solutions. Besides, impressions formed hastily are sometimes difficult to change.

I wanted to assist GySgt E in maintaining his good record. I secured the opinion of legal counsel that GySgt E had acted inappropriately but not illegally. Therefore he should not be made the subject of disciplinary action nor should his reputation be blemished by inference.

My original opinion (formed after reading the SSgt's long letter) changed almost completely. I answered SSgt F's letter and said that I had made GySgt E aware of the charges contained therein and that the evidence did not support any sort of disciplinary action on my part. I further stated that any action he deemed necessary would have to be civil action and should be handled accordingly.

Knowing when to refer a subordinate to other resources — legal, psychological or social — is just as important as knowing how to deploy troops. There is a great temptation to try to be all things to all subordinates. This, we are not able to do as successfully.

QUESTIONS FOR FURTHER CONSIDERATION

1. What is the correlation, if any, between a leaders concept of themselves and their followers concepts of them?

2. What are the most effective and expedient operant conditioning methods to effect positive changes in the relationship between leaders and followers?

3. Is leadership improved most readily by leader or follower behavior changes?

4. Is the "master teacher" concept of education transferable to leadership roles in commercial and service organizations?

5. Can ethics be positively modified to promote leadership capabilities?

6. What criteria should be used to develop a screening device that can eliminate early those unsuited for leadership?

7. What are some of the means that management trainers could provide to humanize leaders rather than dehumanize them?

8. Should the resolution of conflict be initiated by the leader or the follower?

9. Should potential leaders serve an internship under those of proven effectiveness before assuming positions of almost irrevocable authority?

10. Can a democratic relationship derive from autocratic organizations?

part 2
Race Relations

There is no more evil thing in the present world than race prejudice, none at all. I write deliberately—it is the worst single thing in life now. It justifies and holds together more baseness, cruelty, and abomination than any other sort of error in the world.

—*H.G. Wells*[1]

RACE, RACISM, AND RACE RELATIONS

Although outmoded geography books, using color as a criterion, once divided man neatly into five races—white, yellow, brown, black, and red—these arbitrary divisions have no validity, for there is no defensible means by which world populations can be so precisely categorized:

> *Races, however defined, are not fixed entities with precise boundaries. Typologically defined races based on phenotypical likenesses do not correspond to genetic reality. In the light of modern genetics, races can best be defined as interbreeding populations sharing a common gene pool.*[2]

In a consideration of racial matters, a much more practical dictum, and one too often unobserved, is that all living people belong to one and the same species and that the likenesses of our species are much greater than any differences that may be called "racial." Regardless of the manner chosen to define race, it has been found that the individual differences *within* races are greater than the differences *between* races, and that all individuals will vary to some degree in nearly every factor that combines to constitute human beings. Of far greater importance than the variations among humans are the similarities we have which inextricably involve each of us in *all* the implications of the human condition. Acknowledging this truth, many writers believe that incumbent upon us is the collective responsibility of creating a world in which other human beings are accorded the status of persons, and not regarded as things or objects to be exploited.

The term "racism" derives from credence placed on the concept of race; for inherent in that concept is an acceptance of the validity of racial distinctions. Racism, in fact, implies that behavior is determined by race. In scholarly works, the term "scientific racism" is employed to describe a racial interpretation of history, or the belief that peoples of different races have different histories and cultures as a result of their race. However, the vast majority of anthropologists who study both race and culture contend that culture affects race much more than race affects culture. Despite this, in common parlance, the term "racism" connotes discrimination and prejudice. Commenting on this, Whitney M. Young, Jr., defined racism as "the assumption of superiority and the arrogance that goes with it."[3] Almost one hundred years earlier, Benjamin Disraeli had warned: "The difference of race is one of the reasons why I fear war may always exist; because race implies difference, difference implies superiority, and superiority leads to predominance."[4]

In any attempt to understand racism, distinctions need to be made among: (1) institutional structures and personal behavior and the relationship between the two; (2) the variation in both degree and form of expression of individual prejudice; and, (3) the fact that racism is but one form of a larger and more inclusive pattern of ethnocentrism that may be based on any one of a number of factors, many of which are nonracial in character. A review of our culture clearly shows that the historical sources of American race relations are infinitely complex, and there is little doubt that racial bias and discrimination have been built into most American institutions. Ina Corrine Brown concluded, "The United States thus can be called a racist society in that it is racially divided and its whole organization is such as to promote racial distinctions."[5] In this frame of reference, the individual is necessarily a product of institutional racism, but expressions vary from person to person, both in degree and kind. It is also well to remember that what is commonly called racism is in part a segment of the larger problem of ethnic identification, of power and powerlessness, and of the exploitation of the weak by the strong.

To relegate human beings to less than full human status on the basis of their membership in a particular group, whether the group is based on race, class, or religion, is a phenomenon that has become increasingly intolerable to those who are oppressed. To abolish the dilemmas that stem from this racism, institutional arrangements, as well as personal attitudes, must be drastically revised.

What we commonly call *race relations* should be properly understood in the larger context of *human relations* only. Of particular concern should be the expression of attitudes and behavior by people toward others according to their identification as a member of a particular group. The expression of these attitudes and behavioral patterns is not innate but is learned as a part of the cultural process. Because of this, hope that they can be modified positively is justified. Negative group attitudes and destructive group conflicts are less likely to arise when people treat each other as individuals and respond to each other on the basis of individual characteristics and behavior. Students in introductory sociology classes learn that race relations patterns are a part of our *learned* behavior, or of our cultural patterns, and cultural patterns are but the sum and organization of a given people's way of thinking, feeling, and acting. These patterns are not unalterable, however, and with proper processes and patience lend themselves to modification:

> *Today with rapid communication and increased mobility, with the findings of science and the events of history generally made known, people everywhere are becoming aware of the alternatives to old ways. They have access to facts of history and interpretations of science that were previously unknown or unavailable to them. There have thus been opened up to them new conceptions of themselves and of other people. Much of the turmoil of the world today can be traced to the fact that modern communications and mobility have made people everywhere aware of cultural alternatives. All of these things are of the greatest significance in our changing patterns of race relations.*[6]

RACE RELATIONS IN THE MILITARY

The Defense Race Relations Institute (DRRI) at Patrick Air Force Base, Florida, was founded in 1971 under secretary of defense Melvin B. Laird. DRRI was conceived and implemented because abundant evidence underscored the fact that most persons entered the armed forces with insufficient knowledge and appreciation of the culture, experience, and sensitivities of other races to function well in a multiracial environment. When men and women leave civilian life and enter the armed forces, they bring with them their prejudices, myths, and misunderstandings. In speaking to a group of graduates of DRRI, Stanley S. Scott, special assistant to the president, warned those whom he addressed that the task ahead is enormous. He said that building bridges across communication gaps is exceedingly difficult but imperative if "old attitudes are to be converted and biases shattered by the presentation of positive traits."[7] He further reminded them that it is the responsibility of those engaged in race relations work to seek new "answers to questions that refuse to go away."

Organizational problems, placed in bold relief by specific incidents that stemmed from poor race relations, have long plagued, and continue to plague, our military forces. The following brief historical review corroborates Scott's assessment that the task of establishing positive race relations in the military is indeed enormous but also that notable progress has been made.

Shortly after the American Revolution began, blacks, for a time, were barred from the military services despite valiant fighting by many of them in the first battles of that war. When George Washington assumed command of the Continental Army, he placed a ban on the use of black troops because of pressure from the Southern slave-holding colonies, who were afraid that their slaves might mutiny. Later, when 3,000 whites deserted at Valley Forge, this policy was changed, and, once the racial bars were down, white slaveowners in large numbers began to send black men to the Continental Army to serve in their place. When the Revolution ended, 5,000 black soldiers had enlisted and fought commendably in the Continental Army.[8]

Black men were again called upon by Andrew Jackson in the War of 1812 to help save the city of New Orleans. During the Civil War, blacks were permitted to enlist in the Union Army, but their treatment ranged from bare tolerance to outright abuse. About 173,000 blacks served in Union Armies, and, according to war records, black troops took part in 449 engagements and of these 39 were major battles; 37,000, or approximately 21 percent of black troops, were killed. Yet after the Civil War, when black soldiers were sent West to fight, Colonel George A. Custer refused to have them under his command, simply because they were black. How sad it is to reflect on Thomas Paine's words: "A long habit of not thinking a thing wrong, gives it a superficial appearance of being right, and raises at first a formidable outcry in defense of custom."[9] Despite Custer's view, in the 9th and 10th Cavalry units, black soldiers won fourteen Congressional Medals of Honor. Charles Jeffery stated: "The fact that it was during this very period when bigotry became the official national policy with the treacherous Hays Compromise of 1877, a condition that certainly was reflected in the War Department, gives the citations that much more stature."[10] In the Battle of San Juan Hill during the Spanish-American War, black soldiers again distinguished themselves with five Medals of Honor but without historical recognition. Their achievement is all but forgotten in the story of

Theodore Roosevelt and his Rough Riders in most history books.

A further examination of the history of race relations in the military reveals the sordid treatment accorded black troops in World War I who were placed under French command with instructions regarding their treatment. Among those instructions were strong admonitions that other officers were not to eat with black officers nor shake hands with them nor seek to talk with them outside the requirements of military service. Further, blacks were not to be recommended too highly in the presence of white American soldiers, and the native population were given express instructions not to "spoil" the blacks.

The movie *Birth of a Nation,* which in effect was an attempt to justify the Ku Klux Klan, was an example of the state of race relations of this era. However, four regiments of World War I which were awarded the Croix de Guerre (the 369th, the 370th, 371st, and 372nd) were all-black regiments. Of these, the 369th was under fire for 191 consecutive days and was the first allied unit to reach the Rhine.

Despite these evidences of valor and distinction, little change occurred during World War II, in attitudes affecting black men—except that large numbers of blacks were found acceptable for the draft. Dorrie Miller's heroics at Pearl Harbor when he grabbed a machine gun and shot down six Japanese Zeros from the deck of his ship did not even qualify him for the Congressional Medal of Honor.

Poor race relations in the military were certainly not exclusively between whites and blacks, for memory is still vivid of how Japanese civilians were treated during World War II. Initially, Japanese-Americans were locked up in concentration camps and not permitted to serve their country in combat. A letter from Mamie Fujiyama dated June 2, 1974, exemplifies some of the bitterness felt by Japanese-Americans:

> *I was one of the Japanese shut up in the concentration camps during the last war . . . . you are mistaken about our being locked up because we were Japanese. The United States Supreme Court ruled in 1946 as follows: We go on record as affirming that these people were not imprisoned because of their race which is never done in America. Neither were they imprisoned because the Government doubted their loyalty. Such being the case we have always wondered just why we*

> *really were locked up but the Court never explained nor did they ever return the property they "confiscated," I believe that was what they called it then.*[11]

For Japanese-Americans, service in the military began with the Spanish-American War. In 1898, seven Japanese members of the battleship U.S.S. *Maine* died when she was blown up in Havana harbor. Other Japanese-Americans served aboard United States warships in the Battle of Manila Bay on May 1, 1898. At the outbreak of World War II, with the bombing of Pearl Harbor, Japanese-Americans were put in a circumstance which could not help but exacerbate race relations. Many Nisei (second-generation Japanese-Americans) and Kibei (American-born Japanese who grew up in Japan) were of military age at the time of Pearl Harbor. The Navy and the Marine Corps quickly changed the classification of all Nisei to 4C—aliens not subject to military service—thus imposing on them a blanket exemption from the draft. The Nisei deeply resented this discriminatory treatment. In addition, many Nisei already in uniform were released from active service by transfer to the Enlisted Reserve Corps, "for the convenience of the government"—a transparent euphemism for being booted out. No explanation for this action was ever given.

Others of a different racial "classification" have been the victims of similar inequities. In May, 1971, New Yorker Herman Badillo, the first person of Puerto Rican birth to sit with a vote in the United States Congress, presented an appeal on behalf of Puerto Rican people before the House of Representatives:

> *Puerto Ricans have been subject to the draft since 1917 and have served in all wars since World War I. In the current war in Indo-China, for example, some 27,000 Island youths have been inducted into the Armed Forces since the 1964 Gulf of Tonkin Resolution was passed. This represents a higher proportion per population than in any of the 50 states, and almost 300 Island youths have been killed in action. Puerto Ricans from Puerto Rico are required to serve in the Armed Forces of the United States to the same extent as other Americans in all 50 states based upon the fact that Puerto Ricans are American citizens; then the same basis of American citizenship should be used to provide the Puerto*

> *Ricans in Puerto Rico with the same assistance that other Americans in all 50 states receive. This is not the case at the present time.*[12]

American Indians have also had cause for concern in the matter of race relations. Their record in the service of the American armed forces is a long one. In 1778, General George Washington wrote: "I am empowered to employ a body of 400 Indians. If they can be procured upon proper terms . . . . I think they can be of excellent use, as scout and light troops mixed with our own parties." Indians fought for both the Union and the Confederacy in the Civil War as civilian auxiliaries. More than 17,000 Indians fought in World War I, although they were still not entitled to vote and also were not subject to conscription. Yet of that number 6,509 were drafted even without conscription. In the war 331 Indians died, and 262 were wounded in action. It was largely because these men were recognized for remarkable valor that the right to vote was granted to the Indians in 1924, though Indian tribes all over the country considered the gesture as yet another disregard for Indian rights laid down in treaties.

As of 1945, 25,000 Indians had fought in the armed forces. According to 1971 Pentagon statistics, more than 42,500 Indians had served or were serving in the Vietnam war. Countless numbers of Indians who applied for conscientious objector status on the basis of religion were refused the deferments accorded to white Christian petitioners.

Service of Filipinos in the military forces is not exempt from the stigma of negative race relations. During World War I, 25,000 Filipinos enlisted and served in the United States Navy. Other Filipinos served in the National Guard during the same war. By the time of World War II, 4,000 to 5,000 Filipinos were required to register as aliens, ineligible for the draft. In 1941 they protested loudly and petitioned Congress to retain the right to serve. As a result, about 7,000 Filipinos began serving at California's Camp Cooke and Camp Beale—but in an *all-Filipino* regiment.

The Korean war brought to an "official" end the pattern of military segregation, which by then was having pronounced effects on the patriotic dedication of blacks in particular. In that war, for the first time since 1898, black men won Medals of Honor for valor in combat. The

Vietnam war began to put an end to the stigma of inferiority attached to black troops by historians who had remained silent or by propagandists who had fabricated their own "truths." It was during the Vietnam war that blacks were represented for the first time in all branches of the military and in all levels of enlistment and command. However, several incidents in the past few years have offered abundant evidence that the military is not yet immunized against negative race relations and disruptive racial disharmony.

The truth of Thomas Paine's words regarding the defense of custom is painfully apparent in surveying the experiences of the few early black officers—and aspirants to officer rank—and the racial problems that attended the exercise of their rank. A case in point is that of Johnson Whittaker, the third black accepted at West Point:

> *Insult not only followed insult as they did for his predecessors, but injury was laid heavily upon insults. When Whittaker was beaten bloody, unconscious, and tied to his bed on one occasion, no white cadets were charged or punished but Whittaker was court-martialed on the incredible charge of faking the beating to gain sympathy. The outcome was a guilty verdict which forced him to leave the academy.*[13]

Officially condoned discrimination against minorities in military service continued in varying degrees until 1947, when President Harry S. Truman *ordered* the military forces to desegregate. Enough time has elapsed since then to raise an entirely new generation, but the armed forces are still scarred by racial wounds even deeper than before. The Vietnam war served to dramatize some of the inequities in race relations. Advancement in rank for blacks, as well as for members of other minority races, was very slow or nonexistent, pay was low, training in skills was limited, and delegated responsibility was often negligible. Yet despite these negative factors, members of minority races enlisted largely because the military was the only alternative to the unemployment many of them faced. The civil rights movement of the 1960s brought some relief with the battering down of discriminatory laws and some job guarantees, but the decade also saw the Vietnam war claim a disproportionate number of black lives.

Historically, blacks have tended to enlist in combat services rather than in the more technical services. When cumulative hostility to high casualties and other inequities was expressed overtly, those in authority began seriously to reflect upon the wisdom of Daniel Webster's words:

> *If the true spark of religious and civil liberty be kindled, it will burn. Human agency cannot extinguish it. Like the earth's central fire, it may be smothered for a time; the ocean may overwhelm it; mountains may press it down; but its inherent and unconquerable force will heave both the ocean and the land, and at some time or other, in some place or other, the volcano will break out and flame up to heaven.*[14]

Although much of the focus of attention on poor race relations rests with blacks *vis à vis* whites and particularly with inequities in the armed forces, other groups are matters for concern. A study in 1970 revealed that ten million Americans with a Spanish background lived in the continental United States. Paul M. Syscavage and Earl F. Mellor wrote: "In varying degrees these Americans are often beset by many of the same problems blacks face, such as low family income, high unemployment, job discrimination, and lack of adequate education and skills. In addition, they face a language barrier hindering their efforts to achieve economic parity."[15]

Another factor that contributes to lower annual earnings among Spanish-American workers, which, in turn, worsens race relations, is discrimination in hiring and promotion that results in their concentration in low-paying jobs and having higher rates of unemployment. They are found in disproportionately great numbers in such jobs as food service, freight service handler, cashier, and cleaning service workers. Even when Spanish-American men obtained jobs in professional, managerial, and craftsman occupations, their earnings were significantly lower than the earnings of Caucasians in the same occupation group and in the same locality. The destructive effect of these inequities upon race relations cannot be disregarded.

Representative Chet Holifield, Democrat from California, chairman of the House Government Operations Committee, in the *Washington Post* on September 13, 1973, accused the Nixon administration of neglecting Mexican-Americans, Puerto Ricans, and Cubans. Holifield made his comments as representatives of various Spanish-speaking organizations urged the creation of regional offices and an advisory council to make the administration more responsive.[16] Commenting on the character of the Mexican people, Octavio Paz wrote: "A person imprisoned by these schemes is like a plant in a flowerpot too small for it; he cannot grow or mature. This sort of conspiracy avenges itself in a thousand subtle or terrible ways."[17] Spanish-speaking persons take these problems with them into the armed forces.

The armed forces present both a particular *problem* and a particular *promise* in attempts to improve race relations. By their very nature, the armed forces press men and women into close proximity with others from every kind of background. For those who have never known a member of another racial or ethnic group intimately, this proximity often proves to be a positive factor, for everyone discovers that *except for individual differences, there is a common human denominator.* In 1973, recognizing this common factor, and appealing to what is noblest in men, James R. Schlesinger, in his first remarks to the armed services upon being named secretary of defense, posed the race relations problem and the obligation to bring it to an ethical solution:

> *At this point in our nation's history our social fabric is sadly somewhat frayed—reflecting the conflicts over the war in Southeast Asia as well as nagging domestic discontents. We in this Department draw our strength from a healthy and stable society. Restoration of a sense of national unity and purpose must therefore be our highest objective. It is our high obligation to preserve for the future this nation—and to permit the continued flourishing of the free institutions and the social attitudes it represents.*[18]

Chief of staff regulation 15-11, dated August 20, 1973, outlines the establishment, background, mission, composition, direction and control of the Army Race Relations and Equal Opportunity Committee, which was established as a continuing committee on November 2, 1971:

> *The Committee will assure that plans and programs are developed to correct inequities, promote racial harmony and assure all minority personnel equitable opportunity for upward mobility and a full and rewarding career.*[19]

Initiated to redress some of the inequities, the Equal Opportunity program has resulted in a more attentive attitude toward minority needs. But frequent newspaper headlines are evidence that race-relations problems in the military continue. Legislation and executive commitment are resulting in intensive studies and race-relations programs designed in efforts to understand and eliminate race-related confrontations and to find solutions to the problems that are causing them.

In 1973, Donald L. Miller, deputy assistant secretary of Defense for Equal Opportunity, praised the effort being made by this agency:

> *When the Equal Opportunity Officer serves [as he now does] as a full member of the base commander's staff, this is the kind of management we feel has the best chance to move Equal Opportunity from ad hoc efforts and crisis programs to a system of affirmative goal-oriented Equal Opportunity management capable of meeting the root causes of inequality.*[20]

Colonel Ernest Frazier, former director of the Army's Equal Opportunity Program, stated:

> *There has been in the Army a disproportionate number of black cooks, truck drivers, and supply personnel. While in the past there has been overt and covert racism as well as institutionalized racism against blacks, today soldiers are assigned jobs on the basis of what they are qualified to do.*[21]

In addition to training in skills, a new emphasis has been placed on leadership positions for members of minority races. Although there is still room for much improvement, today blacks—but not other minorities—have a much more representative posture in the armed forces than they have had heretofore.

Clearly, much remains to be done. In 1973, the Army had nine black generals, including two who each commanded one of that service's thirteen divisions. The Air Force had two black generals and the Navy had one admiral. The Marine Corps had no black general officers. Racial inequities are at once apparent, however, when a percentage comparison is made from statistics available in March, 1973. During this same year, the Army had 15.1 percent black soldiers, and only 3.9 percent held officer rank. All other minority races combined comprised 1 percent of the Army, with only .4 percent holding the rank of officer. The Navy had 5.7 percent black members, and 4.1 percent other minorities; of these, only .9 percent black were officers and .4 percent other minorities were officers. In the Marine Corps with a black enlistment of 12.5 percent and other minorities 1.3 percent, only 1.5 percent of the blacks were officers and only .5 percent of other minorities held officer rank. In the Air Force, 10.8 percent of the total forces were black and 7.6 percent were other minorities. Of these totals, only 1.7 percent of the blacks were officers and .6 percent of others held officer status.

The effect of such racial imbalances in leadership roles is apparent in the formation of negative attitudes that result in racial confrontations and conflict.[22]

SPECIFIC COMPLAINTS AND CONFRONTATIONS

Despite an inclination on the part of some members of the military to continue the stance of an ostrich with its head in the sand, explosive incidents within the armed forces have forced those in authority to review their thinking and to seek more effective measures to deal with race relations problems. Some examples of racial conflicts that forced this review are summarized below:

A series of fights flared throughout the night of October 12, 1972, aboard the U.S.S. *Kitty Hawk* in the Gulf of Tonkin. Forty-six persons were injured and twenty-six were arrested. As a result, perjury charges were brought against Michael Laurie, a white, because of his testimony against a black shipmate, Cleveland Mallory. Mallory was convicted and given a bad conduct discharge on the basis of Laurie's testimony. Three days after the NAACP filed a federal court suit against the Navy charging Laurie had lied, Mallory's conviction was reversed by a Naval review authority on the grounds of insufficiency of evidence and Mallory later was given an honorable discharge.[23]

On September 7, 1972, six black and four white Marines battled each other aboard the U.S.S. *Sumter,* an amphibious assault ship. A little over a month later, on October 16, 1972, four whites were injured and eleven blacks were jailed following a brawl on the fleet oiler U.S.S. *Hassayampa.* In November, thirty-two blacks and one white were locked in individual cells after a confrontation at the Navy's correctional center in Norfolk, Virginia. In the same month, four whites and one black were slightly injured as more than one hundred sailors engaged in a free-for-all on Midway Island. In another incident that November, a number of black sailors staged a sit-in aboard the U.S.S. *Constellation* to dramatize their grievances and protest rumored discriminatory discharges.

On November 12, 1973, the *San Antonio Light* reported that Rudy García, a Mexican-American employee at Kelly Air Force Base, had filed suit charging Air Force officials with discriminatory employment opportunities at the air base "due to their national origin."[24] The plaintiff alleged that, "although Mexican-Americans constitute more than 50 percent of the population of San Antonio, the percentage of Mexican-Americans in higher-paying job classifications at the base is insubstantial."[25]

On November 30, 1973, the *Washington Star-News* reported that a black Army colonel had been rejected for service as a military attaché in Chile by the United States Military Advisory Group (Milgroup) there, and had also been turned down for a similar position in Bogotá, Colombia. It was reported that the colonel's race had not influenced either decision. On December 1, 1973, however, the Defense Department acknowledged that Navy Captain R.E. Davis, Chief of Milgroup in Chile had advised Washington that "a black colonel would not be acceptable in Chile because of his race."[27] In actuality, Chile was not consulted about Colonel T.M. Gafford's assignment. Colonel Gafford did not get the assignment because, as Pentagon sources allege, he lacked the proper occupational specialty. Such situations, even if rectified, reflect negatively on military race relations.

On December 4, 1973, the *New York Times* reported that "three outbursts of racial violence erupted among American troops in Tongduckon, South Korea."[28] These outbreaks evidently touched off similar encounters among American forces elsewhere in South Korea. In the aftermath of one of the 1973 incidents, of the thirty-eight men involved, mostly blacks, charges against four were dropped; twenty-seven were given discharges instead of facing court-martial or were found unfit for military service because of past involvements in racial trouble; two are awaiting possible court-martial; one was given minor punishment; two were given bad-conduct discharges; and two were sent to military prisons in the United States. The Washington, D.C., headquarters of the American Civil Liberties Union sent an attorney when it was reported that but one single Army captain was assigned to defend twenty-two of the cases.

On September 7, 1973, *Stars and Stripes* reported that Staff Sgt. Ernest Tabb, a former prisoner of war in Vietnam, complained bitterly to the Augusta City, Georgia, City Council that he did not receive the recognition given to other POWS because he was a black man. One councilman, Hugh W. Cross, commented, "It is certainly my opinion that his bitterness may be justified."[29]

On August 28, 1973, the *Washington Star* stated in an article from Bamberg, Germany, that, according to a United States Army spokesman, military police had arrested fifteen black soldiers after firebombing incidents and racial clashes at Warner barracks.[30] Five soldiers were injured in this racial confrontation and one was hospitalized with a broken jaw.

> *The* Overseas Weekly *reported in March, 1974, from Naples, that ten black sailors and two white sailors have been charged with rioting and assault on the Flagship U.S.S.* Little Rock *of the U.S. 6th Fleet during the Mid-East crisis in November of last year. The ten blacks claimed they were victims of racial discrimination and could not expect a fair trial under existing circumstances. The sailors claimed the flack started after months of racial tension aboard with numerous instances of "institutionalized racism" by the ship's officers.*[31]

Anthony Griggs wrote:

> *Aside from residual discriminatory treatment of American minority soldiers, other critics point to what they call "racist" and "genocidal" intentions behind the reason and direction of the Vietnam War. This accusation — levied as much by sectors of the domestic society as by the minority enlisted men — portrays the whole of the military complex as a racist institution both in content and intent.*[31]

THEORIES AND PROGRAMS FOR BETTER RACE RELATIONS

Such incidents as those above dramatize some of the race-relations problems suffered by the military. Major Jon M. Samuels captured the effects of these altercations:

> *Millions ponder the national security implications of this or that incident. Congressional committees investigate and pundits speculate. Buffeted by these pressures, the service professional must contend simultaneously with angry minorities, an inveterate institutionalized resistance to change, and pious demands by "above-the-battle" superiors for the immediate resolution of these "embarrassing" confrontations.*[33]

A number of theories have been advanced and programs initiated in an attempt to cope with the race relations problem. Despite the accusation of Curtis R. Smothers, former Deputy Assistant Secretary of Defense, in May, 1973, that there "is a damn serious failure to move forward with credibility in relating to blacks, women, and other minorities,"[34] much that is commendable is being done and new programs continue to be formulated and implemented.

Commenting on these efforts as they pertain to the Army, Colonel Frazier said, "Progress in race relations is being made throughout the Army but the potential for confrontations and violence still remains high."[35] One of the most positive factors reported by Colonel Frazier after he visited units in the United States, Europe, and the Pacific, was that senior commanders seemed to have a greater awareness of racial problems and were generally committed to resolving them. Discerning one problem as lack of minority leadership, the Army and the Air Force appear to be accelerating their efforts to promote qualified minority members to command posts. Army officials believe that, especially in Europe, this attitude has assisted in easing racial tensions. Race-relations training has been instituted, and these training efforts begin with those who are in command positions, which is an indication to all that the Army "means business."

The race-relations training makes a genuine attempt to identify and correct problems that interfere with racial harmony. One approach is to provide both white and minority members with deep insights into the genesis of the race problem. Training officials are having considerably more success with racial awareness programs conducted in three intensive six-hour sessions instead of spreading them out over several months. At Fort Sill, Oklahoma, it was determined that race-relations seminars containing no more than twenty-five participants were most effective; however, motion pictures and lectures can serve a larger audience.

The establishment of DRRI as an interservice school to train instructors to assist in such efforts as those described above, began at Patrick Air Force Base, Florida, in 1971. By June, 1975, DRRI had placed nearly 3,500 trained race-relations instructors in military units around the world. These graduates conduct Department of Defense seminars in race relations in an organized effort to combat the military's racial troubles.[36]

The Navy also has embarked upon a massive race relations education program. *The Navy Times* reported on March 21, 1973, that naval commands in San Diego, Newport, Norfolk, and eventually Hawaii are joining the Memphis Human Resources Department in the program designed to provide additional men to conduct these seminars. Across the United States, these commands will initiate the first Race Awareness Facilitator Training (RAFT) system drawing future facilitators from every base.[38] The RAFT program recruits students on a voluntary basis, trains them, and then returns them to their respective commands. The objective is to help Navy personnel understand their personal worth and racial dignity. These units are supposed to be the foundation of the Navy's race-relations education program.

Recently, when Okinawa-based Marines were plagued with racial tension, a program was developed to enable Marines to take the lead in improving race relations. One purpose of this program was to improve relations not only between Marines and Marines but also between Marines and civilians. The human relations program implemented was divided into three phases — orientation, discussion, and action:

> *The human relations program on Okinawa involved 22 unit discussion leaders . . . the discussion leaders were sent to three weeks of training to learn how to control discussion groups, bring out different views, and to get individuals to reevaluate their own thoughts.*[38]

Jerry Pace, one of the human relations instructors, said that, according to critiques given at the end of the program, most negative attitudes seemed to have dissipated by the end of the course.

Also in 1973, behind closed doors at Homestead Air Force Base, men and women in uniform were playing serious "prejudice games" in an effort to improve race relations. These simulations of real-life situations can modify negative behavior and increase the awareness that many verbal terms which are used (and that most people are not even aware of) can cause serious race-relations problems. The following is an example of these games: Without offering solutions, instructors ask mixed groups to list things that they believe contribute most to impersonal or institutional racism. In one session, answers included insufficient low-income housing, restrictive zoning, federal housing policies of the 1950s, blockbusting by realtors, and urban renewal practices. In these games, the transition is then made from the general to particular or personal racism. In another game, students were told to put down the first word that came to mind in response to five key association words: Indian, Mexican, Asian, white, and black. Answers came back like — oil, farmer, Jap, person, and nigger. Many participants looked at their own prejudices for the first time.[39] Lieutenant Bruce Bradburn, one of the staff of five at Homestead, explained that no attempt was being made to absorb or eradicate color. In fact, the emphasis was on preserving separate identities for those who choose to retain them but at the same time to treat each person with the dignity he deserves simply because he has been created human.

Racial sensitivity has become big business in the military establishment. By mandate from the Department of Defense, every man and woman in military service must receive military-oriented training in race relations. Evaluation of programs already implemented has shown the wisdom of this decision. Establishing meaningful communication and sharing pent-up resentment and hostility are resulting in new insights about self and others that can be beneficial in establishing more positive race relations in the military.

What is unfamiliar to us — we do not understand; what we do not understand — we fear; what we fear — we hate.

> *One needs to hear Job lift his question into the wind; it is, after all, every man's question at some time. One needs to stand by Oedipus and hold the knife of his most terrible resolution. One needs to come out of his own Hell with Dante and to hear that voice of joy hailing the sight of his own stars returned-to. One needs to run with Falstaff, roaring in his own appetites and weeping into his own pathos. What one learns from these voices in his own humanity.*[40]

Race Relations at Base X

Captain Jesse L. Dansby

Air Force Logistic Command,
Equal Opportunity and Treatment Human Relations Officer
Wright-Patterson Air Force Base, Ohio

METHODOLOGY

Extensive formal interviews with 170 servicemen of all ranks and ethnic groups were conducted over a five-month period. The interview schedule consisted of one hundred questions, many of them openended. In addition, several informal interview sessions with dependent wives were held, and many servicemen were interviewed on an informal basis. Servicemen were also observed on the job and in off-duty situations.

GENERAL FINDINGS

Although it is difficult to determine accurately the mood or atmosphere among racial or ethnic groups at any military installation at any given time, the overall relationship between black and white servicemen at Base X may be labeled fairly good. This does not mean, however, that a major racial conflagration is unlikely to occur nor that minor racial incidents might not occur. A single incident or precipitating event could lead to major problems. The overall atmosphere at the base is neither tense nor harmonious. Blacks and whites tolerate each other; some are friendly on and off the job. On the job, the two groups work well together; off the job they tend to separate. Racial separation, self-imposed, exists. It may be viewed at the NCO club, in the mess hall, and in the housing arrangements in the barracks.

Three overriding facts were evident from the research: (1) a white "backlash"; (2) a black perception of discrimination and racism; and, (3) extreme race consciousness by black servicemen. Many whites believe that the U.S. Air Force is bending over backwards, giving blacks more than an equal opportunity. Furthermore, whites feel that blacks are taking advantage of the situation, claiming prejudice and discrimination when anything goes against them and that white supervisors are bowing to black pressure. Nevertheless, blacks believe that discrimination and prejudice still exist in the Air Force, although they find it subtle and difficult to detect. They feel the discrimination is both individual and institutional. Few blacks were able to pinpoint exact cases of discrimination but rather expressed it as, "It's in the air," or, "You can just feel it," or "You can tell the way a person speaks to you." As one black A1C stated, "If a person were blind and black, he could see that they [white personnel] are prejudiced just in the way they talk to you — as if you're some kind of beast, a dog or animal." Throughout the formal and informal interviews and personal observations, the black serviceman was found to be extremely conscious of his race, viewing himself as a black man first with his own history, heritage, and culture, and secondly as a serviceman. Often the white could not understand the black serviceman's feelings on racial issues, and this caused problems and complications.

The black servicemen were found to be much more outspoken and loquacious than their white counterparts. Whites were more hesitant to discuss their racial feelings than the blacks, obviously fearing that they would be labeled as bigots and prejudiced — as many of them proved to be. Conversely, blacks were not afraid to discuss their inner feelings on race relations on the base and about the whites on the base. Some blacks came across as racist as did certain whites, disliking whites, disliking taking orders from whites, and disliking interracial dating. Another major finding, serendipitous in nature, was the lack of knowledge by the officers and senior NCOs about the various equal-opportunity and social-action programs and the racial situation on the base. Not only the pilots but squadron commanders and supervisory personnel were completely out of touch with reality.

Ignorance among all ethnic groups is the major problem in race relations in the Air Force. Blacks, whites, Mexican-Americans, and Orientals do not understand other ethnic groups and seem unwilling to learn about other cultures. However, the problem is primarily a dilemma for whites, since they are the dominant group in the Air Force, both numerically and in the power structure. For example, one sergeant explained to me that he loathed blacks because of his experiences in Vietnam. A month later he expressed that he had altered his views after watching a television documentary on Dr. Martin Luther King, Jr. He stated that his ignorance (his own word) of the problems the black American faced – and faces – had led him to a narrow-minded opinion.

Regarding Air Force policy, whites do not understand the reasons for the new thrust in equal opportunity, and blacks do not trust decisions made from the establishment, with a definite credibility gap the result. Although not all problems will be solved by broadening an individual's horizons, increasing awareness of other ethnic groups, or interacting with all types of individuals, programs in these areas will certainly eradicate many of the factors inhibiting harmonious race relations.

GENERAL PROBLEMS AT BASE X

Dissatisfaction with Personal-appearance Regulations

Both blacks and whites believe that a double standard exists in implementing the regulations, with the opposite group not only flaunting the regulation but receiving preferential treatment. A white staff sergeant complained, "Black airmen seem to get away with a different personal grooming standard than we do; you see some awfully bushy heads running." A black A1C remarked, "Whites' hair is longer, but he can grease it down, while ours is curly and sticks out." Whites, unaware of a skin disease inherent in blacks and in some whites, also complained that blacks were permitted to raise beards without valid medical reasons. Both racial groups, especially the young members, complained that their short hair was not in keeping with the times and prevented them from expressing their identity.

Lack of Knowledge of Equal-opportunity Programs.

Those who were aware of the programs and policies were skeptical of their success. Few servicemen know anything definitive about the Social

Actions Office owing to a lack of publicity and communication. A white colonel was unfamiliar with the Air Force equal-opportunity programs because he was not "directly involved with them." A black sergeant realized that the programs existed, but felt that they were valueless since "nothing ever results from them." In addition, the personnel in the Social Actions Office were unsure of their role on the base other than conducting the racial-awareness seminars.

Understaffing of the Social Actions Office.

With only four assigned slots, the office is seriously undermanned. Sending the drug-abuse officer and the NCO to their respective schools simultaneously depleted the staff by half. Furthermore, with no guidelines for its function, the Social Actions office has had problems determining its overall role and position on the base. By becoming the focal point for various programs — the Martin Luther King Memorial Service and luncheon, and the People's Expo — the office has been tied up with peripheral duties, so that its primary function — the conducting of racial awareness seminars — has not been carried out. Certainly the sending of the equal-opportunity officer to the Defense Race Relations Institute, followed by the NCO at a later date, was not a wise maneuver.

Complaints by Minority-Group Servicemen

1. Lack of desired products at the base exchange and mess hall.
2. Lack of desired entertainment at the NCO club.
3. Unequal punishment at squadron level. Minority-group members were punished more severely than whites for similar offenses.
4. Supervisory harassment relating to personal appearance, tardiness, and condescending attitude.
5. Discrimination in promotion, especially on Airmen Performance Reports (APR).
6. Absence of black personnel at officer level and top enlisted ranks.

Complaints by White Servicemen

1. The Air Force, oversensitive to black pressures and demands, has overreacted in its equal-opportunity programs, giving more than equal opportunity to the black serviceman. Remarks ranged from the captain who felt that the Air Force was pushing "too hard," to the major

who believed that the Air Force was "going overboard to prevent racial polarization — they are more concerned about racial problems than they should be," to the white sergeant who believed, "Whites must treat them [blacks] with kid gloves," to the A1C who felt that the Air Force was letting blacks do what they want "by letting them grow beards while whites cut their throats shaving."

2. Excess of soul music at the NCO club.

3. Discrimination in work distribution with whites receiving the heavier work loads.

4. Unequal judicial punishment at the squadron level.

Because of the increased interest in equal opportunity and the emphasized importance of effective human relations, many servicemen have overreacted to race relations. Instead of attempting to improve relations in the long run by implementing programs and policies, the Air Force and many of its personnel have concentrated on stopgap measures aimed at preventing race riots. Furthermore, supervisors know that equal opportunity is a hot issue and that they must at least pay it lip service — and oftentimes that is all they give it. As a white captain stated, "I think that the reason equal opportunity is pushed is not for equal opportunity's sake but for the sake of the military, to prevent racial unrest and racial tensions. The primary function is to have the military function smoothly, and secondarily, to promote equal opportunity for blacks." The result, unfortunately, is that blacks often receive differential treatment, particularly from supervisors who fear that blacks will protest prejudice or discrimination if they make a decision against a black serviceman. Blacks as well as whites are aware of this situation, and neither group likes it. Blacks do not want to be treated differently; they simply want fair treatment and justice.

Case Report

The first overt expression of dissatisfaction was the taking over of the base recreation center by a group of some sixty-five black airmen. The commander seriously entertained the idea of forcibly removing them. However, he was advised that the incident would remain relatively peaceful and might be negotiable if he did not overreact. Therefore, no attempt was made to forcibly evict the demonstrators. Negotiations were conducted, with the result that the demonstrators left the recreation center twenty-four hours later. The demonstrators seemed to be leaderless; however, four members of the group were selected to represent them and conduct further discussion with the installation commander. This discussion was conducted after the demonstrators had left the facility, at which time the grievances were made known.

During the inquiry that followed, over 300 blacks were interviewed. In addition, an estimated 600 other airmen, NCOs, and civilian supervisors were interviewed. The interviews revealed that there were many other grievances in addition to those that had originally surfaced. The interviews and observations revealed that:

1. The demonstration had been spontaneously planned the night before. The demonstration was planned to be peaceful; however, the participants were prepared for violence and the consequences of forceful apprehension if that followed their actions.

2. Black airmen had conducted several meetings and were organizing ethnic cultural and study groups. An abbreviated list of the grievances is outlined below:

a.) The situation concerning discrimination against blacks on base and by the local community which had existed during an evaluation of attitudes and feelings of personnel conducted earlier in the year was still prevalent. No perceptible improvement had occurred, and there was no way a black airman could find off-base recreation — social or religious diversions in the local community.

b.) Because of the conditions stated above, the black airmen believed that the tour of duty at this installation should have been shortened.

c.) Because of the conditions in a.) above, the black airmen were forced into driving a certain distance in order to obtain recreation and relaxation. Such mobility increased their exposure to accidents, traffic violations, and mechanical difficulties, which, in turn, caused other incidents, such as reporting late for duty. Some late reporting resulted in disciplinary action, in spite of the commander's policy of leniency if the authorities at the base were properly notified in advance of reporting times.

d.) The black airmen believed that they were being administered an inordinately large percentage of the nonjudicial punishments, as well as being unduly harassed by the Air Police.

e.) The black airmen believed that passive discrimination was being practiced against them by the local managers of on-base hiring activities, such as the base-exchange, bowling-alley, and dining-hall contractors. Applications for employment had been submitted by blacks but had mysteriously disappeared from files, and only token employment for blacks had occurred in these agencies.

f.) The black airmen believed that the Human Relations Council was ineffective and not representative.

g.) A large percentage of the black airmen stated that when they brought problems to their supervisors, nothing was done about them.

h.) The black airmen complained of lack of common courtesy in the dormitories on the part of maids, civil-engineering employees, and other inspection personnel, who entered their rooms without knocking.

i.) There were complaints about continual supervisor harassment of blacks with authorized shaving excuses because of pseudofolliculitis (ingrown hairs), as well as the difficulty in securing medical waivers for not shaving.

j.) The airmen spoke of extensive use of and threats to use the Record of Individual Counseling, with the connotation that all such counseling is derogatory in nature.

k.) There were daily occurrences of the words "nigger" and "boy" directed at the black airmen.

l.) The airmen believed that praise was not given when earned but that supervisors were quick to criticize.

m.) Black women were unable to obtain satisfactory hairstyling service in the base beauty shop.

3. Virtually all the grievances were substantiated by the inquiry team to one degree or another. In addition, it was determined that a serious lack of communication existed between first-term airmen (black and white) and their supervisors at all levels.

Minimization of the individual's problems and failure to realize the importance of providing a solution were common. Little evidence of aggressive problem-solving, follow-up action on the part of top- and middle-level supervisors was ascertained. Some supervisors appeared reluctant to take an individual's problem to the various councils that existed for such problems, and in many cases they were unaware of the procedures to do so. Most of the first-term airmen believed that the commander's "open-door policy" existed in theory only and that they had effectively been deterred from pursuing it because of their personal fear of the command atmosphere into which they would be entering. Their concern over the consequences of not being completely proper in speech, dress, and military bearing had kept many from taking their problem to a higher level.

4. There were positive indications of an imbalance between black and white airmen being assigned to remote duty.

Conclusions

1. This installation is unique in that there is absolutely no black community nor a section of town for off-duty recreation and relaxation by the black airmen. Discrimination against the blacks by the local community was openly practiced.

2. The Human Relations Council had been relatively ineffective.

3. The serious lack of communication that existed between the young airmen, especially minority groups and supervisors, precluded the establishment of a satisfactory dialogue between them and base authorities.

4. The number and nature of grievances made by the black airmen, as well as the local civilian-community situation, contributed to the decision to conduct the demonstration.

5. The black airmen had experienced undue harassment both on and off duty.

6. Additional recreational and relaxation facilities and activities were needed.

7. Racism against the blacks had been practiced on base.

8. The percentage of black first-term to black career airmen at this base was considered to be inordinately high. The civilian-community situation and the relative isolation of the base presented a much greater problem to first-term airmen.

Recommendations

Actions and guidance that had been developed and implemented at headquarters-command level were begun and closely monitored at this base. It was recommended that:

1. Positive action at all levels be initiated to persuade the civilian community in the area to implement existing nondiscriminatory and fair-treatment laws "in fact and in spirit."
2. Military regulations against discrimination be rigidly enforced.
3. Personnel assignment policies for single, first-term airmen be reexamined with a view toward minimizing the number of black, first-term personnel assigned to this installation.
4. The commander be provided specific recommendations for elimination or reduction of all irritants.
5. A race-relations awareness course geared first toward top management and supervisory levels be instituted, the class to be made up of peers.
6. A full-time, equal-opportunity officer be assigned to the base. Criteria, duties, and responsibilities of this individual are outlined below:

The individual appointed to this position must be perceptive and sensitive to social problems and possess the desire to serve in the human relations field. He must also possess a temperament and personality that will enable him to relate to people. He must have a recognized capability for solving problems without being dominated by executive or administrative authority. The EOO will serve as the commander's personal representative on all matters associated with the Equal Opportunity Program. He will insure that the program within each unit is aggressive and positive. He should be permitted easy access to, and maintain close liaison with, the senior installation commander. His office will serve as the focal point for matters pertaining to equal and fair treatment of military personnel. He will also assist the commander in completing the semi-annual, equal-opportunity status report. A sample of this report is given below:

> *1. How would you rate the racial climate on your base? (Indicate reason for rating):*
> *a. Excellent b. Good c. Fair d. Bad e. Serious*
> *2. How would you rate the racial climate off-base? (Indicate reason for rating):*
> *a. Excellent b. Good c. Fair d. Bad e. Serious*

3. Is there a unit in your organization with a disproportionate concentration of only one racial/ethnic group? If your answer is yes, are you aware of any adverse effect this may have on other members of your organization? If necessary, what local action has been taken to create a better balance?

4. Is there any overt evidence of the so-called "white backlash" occurring on your base? If so, how is it manifested and how has it been managed?

5. Make an analysis of minority-group job assignments at supervisory level (rank versus duty and responsibility). What is your assessment? If problems are discovered, outline corrective action taken.

6. What local procedures do you have for investigating and reporting cases of alleged discrimination? What method do you employ to publicize results of sanctions, etc., in cases where discrimination is substantiated?

7. Report summary of discrimination complaints and action taken on each case.

8. Make an analysis of where all of your people are living both on and off base. Are there indications of de facto segregation, polarization, discrimination? Outline specific corrective action taken in each case.

9. How many people are on your Housing Referral (HRO) *staff? What action have you taken to insure that the staff is sensitive to the problems of minority-group members in locating economy housing? Do you have minority ethnic-group representation in your* HRO *staff? What procedures have been established to insure a strong positive working relationship between the* EOO *and* HRO?

10. Have you read and are you and key staff members familiar with regulations pertaining to EOO *and Race Relations — specifically,* AFR *35-78 and* AFR *50-26?*

11. What action have you taken to formalize (administrative and documentation) the Equal Opportunity and Race Relations programs? What local goals and objectives have you developed?

12. What procedures have been established to educate personnel on the Military Justice System, i.e., Article 15, 39-12, 39-10, discharge, other than honorable and its disadvantages, etc.?

13. What are the AFR *39-12, Article 15, and Court Martial case rates for your installation? What are the comparative statistics between minority groups and the white population? How often do you review these data? Are the rates greater for your minority groups than for your white population? If so, provide candid appraisal of the reasons for the disparity and outline any local action taken to insure effective and fair management of the military-justice system.*

14. Are any subtle discriminatory practices discernible in the operation of recreational activities?

15. What steps have you taken to insure that base exchange barbers/beauticians are qualified to furnish haircuts and hairstyles for minority male and female patrons?

16. Is the Human Relations Council composed along the guidelines published in the Headquarters Human Relations/Equal Opportunity Meeting minutes and the Headquarters Chief of Staff letter dated 19 April 1972? Provide data on age, race, grade, and sex of your council membership.

17. Do you review the Human Relations/Equal Opportunity minutes? How do you indicate your actions taken? When and how are the council members informed of your actions?

18. How is the council being used to further education and awareness of EO/HR *programs?*

19. What media do you utilize to orient all personnel of your base community of the Air Force's and your policy regarding equal opportunity and fair treatment?

20. What vehicle/forum do you employ to insure newly arrived personnel receive indoctrination on local Equal Opportunity and Race Relations policies?

21. Have you attended the Race Relations course conducted on your base? What has been done to achieve an ethnic mix in your classes? What provisions have been made for commanders, key staff members, first sergeants, security police, supervisors of large groups of personnel, club custodians, and other key supervisors to attend the Race Relations course as soon as possible?

22. What procedures has your Chief of Security Police established to coordinate with local civil law enforcement officers to assure uniform treatment of all military personnel in the civilian community?

23. Have law enforcement personnel received training to enhance effective human relations?

24. Please forward a copy of your personalized EO *policy statement.*

25. Do you have an open-door policy? If so, what are the operating procedures? Is this same procedure practiced by other commanders of your wing? Do you seek feedback on the issues which concern your military community?

26. What is the commanders' personal involvement in exercising direct supervision over the EO/RR program?

7. An investigator system such as the one outlined below should be developed:

OVERVIEW

Before proceeding with an investigation, certain facts must be determined. The complaint must be identified, as well as the nature of the discrimination, that is, complaints of discrimination owing to disciplinary actions and/or discharge, a complaint of discrimination on performance reports, housing, and so on. The subject is too broad to identify readily all the potential causes of discrimination, but in each case these points must be known before an effective investigation can be accomplished.

GUIDELINES

Person Receiving Complaint.

The person receiving the complaint must adhere to the following steps:

1. Obtain the name, address (military where possible), title, and grade of complainant.

2. Identify and describe the nature of the alleged discrimination, including the date and time of the occurrence.

3. Inform the complainant when he can reasonably be expected to receive a reply. The nature and complexity of the complaint should be considered in making such a commitment. In any case, a concerted effort must be made to give an interim reply within five working days or less.

Person Assigned to Investigate Complaint

At the earliest opportunity, the investigator should interview the complainant. Before doing so, however, he should do the following:

1. Review the complainant's charges or allegations.

2. Identify the type of complaint and make certain that each charge or allegation is clear.

Actual investigation of Complaint

It is imperative that the investigator maintain impartiality and proceed with the investigation in a logical and orderly manner. Specifically, he should

1. Investigate the complaint at the earliest possible date.

2. Win the confidence of the complainant, as well as any other person he interviews, by remaining receptive to matters pointing up the need for correction or improvement in the EO *program even when he does not agree and even though the comments may not be directly related to the case.*

Processing the Complaint

The equal-opportunity officer should thoroughly review the results of the investigation and submit a summary of his findings and recommendations to the commander through the social-action officer. The staff judge advocate should participate in the review of the written summary of all facts obtained by the investigating officer. The stated purpose of this review is to assure that the report is adequate, factual, and responsive to the allegations of discrimination made by the complainant.

SUMMARY OF PERTINENT FACTS

A short review of the more significant information should be presented in this section. Be careful not to generalize or develop unrelated facts.

FINDINGS

All relevant findings should be noted in short but precise statements.

RECOMMENDATIONS

Recommendations should be presented based on the findings and any other pertinent data. It is important to remain on the subject and to avoid generalizations.

8. An effective Human Relations Council be established: The council will be composed of a cross section of assigned and tenant

personnel, including civilian personnel, and will be multiracial in composition, with both male and female representatives. The bulk of the membership should be from junior officers down through the enlisted grades, including airmen in grades E3 and E4; lower-grade airmen should comprise at least 50 percent of the membership. The council should have at least twelve members, with not more than 25 percent from the senior NCO/officer structure. All personnel and their dependents should be encouraged to utilize the Human Relations Council to air grievances, offer recommendations, and so on. Council actions and publicity must insure that there is no implication of usurping or replacing a member's prerogative to submit formal complaints as regulations provide. The council will meet at least monthly; minutes will be recorded and copies posted conspicuously. To insure constant activity and new ideas, a change in membership should be considered at least annually; however, it is not recommended that the entire membership be changed at the same time. The functions of the Human Relations Council are:

a.) To provide a forum to inquire into and make recommendations with respect to complaints and grievances arising from official or unofficial activities.

b.) To research, analyze, and recommend appropriate preventive measures to commanders concerning implementation of the principle of equal opportunity and to provide a vigorous program of education through every available medium.

c.) To serve as a sounding board for providing the commander information about the degree of racial hostility or tension.

d.) To take an active part in social-action problem identification and solving.

e.) To explore and develop new ideas, concepts, and programs that will lead to a more harmonious relationship between the races.

9. A general awareness and firm action program be developed. The senior installation commanders will:

a.) When reviewing the Human Relations and Equal Opportunity minutes, make specific references as appropriate to include actions to be taken, his concurrence, nonconcurrence, and rationale.

b.) Lend full support to personnel he appoints to investigate complaints, grievances, and so on.

c.) Insure that administrative discharges are not used as a substitute for actions that could otherwise be corrected through sound leadership and management practices.

d.) Encourage and practice a true open-door policy at each echelon of command to facilitate the identification and remedy of unequal treatment, whether real or imagined. This policy must be flexible enough to accommodate personnel who work irregular hours, such as shift workers.

e.) Establish realistic equal-opportunity and human relations goals and objectives for his installation.

f.) Establish procedures to insure that the Social Action Office or EOO is apprised of formal complaints submitted under other channels when it is clearly evident that the matter is related to EO or human relations.

g.) Insure that during initial in-processing newly assigned personnel are advised of local policies regarding equal opportunity.

h.) Insure that one person in every squadron or detachment, other than the commander, will act as the EO liaison officer or NCO, as appropriate, who will work closely with the parent organization equal-opportunity officer.

i.) Insure greater participation of all groups in planning recreation and entertainment programs to assure broad appeal. Commanders will insure that applicable councils represent the clientele they serve.

j.) Insure that military justice will be firmly and impartially applied when circumstances warrant such action. Race or minority status will not enter any consideration for punishment. There must be no possibility that disciplinary action is being taken, or not taken, for any reason other than the facts warranting that action. Commanders will publicize punishments under Article 15, UCMJ/Court Martial through such vehicles as unit bulletin boards, daily bulletins, and newspapers, after removal of the name of the member concerned. Statistics will be maintained by SJA on the number of individuals administered punishment under the UCMJ to include charges, convictions, and punishment, and will be broken out by rank and race, that is, blacks, whites, Spanish-Americans, and so on.

k.) Establish positive procedures, such as assigning control numbers to applicants in order of receipt, to insure that employment practices by nonappropriated fund activities and on-base operated concessions, such as banks, credit unions, and base exchanges, provide equal opportunity and treatment of all military and civilian personnel irrespective of race, sex, color, or national origin, consistent with physical capabilities and occupational qualifications.

10. A management training program be developed to expose managers at all levels to the philosophy and techniques of modern-day management principles such as those of Maslow, Hersey-Blanchard, and Gellerman. The general purpose of this program will be to develop more enlightened, better-prepared, and better-organized managers. Course material will at the minimum include the following:

a.) Behavioral approach to management.
b.) Human relations movement.
c.) Motivation and behavior.
d.) Supervisory styles.
e.) Management philosophies.
f.) Communication.
g.) Motivating environment.
h.) Leader behavior.
i.) Determination of effectiveness.
j.) Management for organizational effectiveness.

11. A Career Advancement Program (CAP) be developed. This will be a free-of-charge, on-duty educational program for personnel who need reinforcement in the basic communicative skills — reading, writing, mathematics, and so on. These courses will be taught on an individual basis. Class size will be kept to such that will allow for personal attention of the instructor to each student to identify the specific area in which he indicates a weakness.

The program as indicated above is but a scratch on the surface; however, it appears that we might be headed in the right direction. Since the institution of these programs, there has been an indication of improved racial harmony. But who knows what will come with tomorrow's dawn?

QUESTIONS FOR FURTHER CONSIDERATION

1. Would human-relations workshops as a preliminary requisite for all inductees into the armed forces be more beneficial than such exposure later in the training period?

2. Are military personnel more responsive to civilian human-relations facilitators without rank than to those selected and trained within the military?

3. Should tests be devised and administered to recruits to determine their "race-relations suitability" for service in the armed forces?

4. What rewards and punishments are most effective in establishing and maintaining good race relations?

5. Are structural or institutional changes in the armed forces possible that will alleviate the race-relations problem without undermining the discipline necessary for military personnel to perform their functions?

6. Should each branch of the armed forces operate independently in remedying poor race relations, or should a common set of methods and procedures be established?

7. What is the implication of the Gafford affair? (see page 54)

8. What changes have occurred in race-relations training in the military since 1974?

9. Should race-relations training be integrated into all military leadership training units?

10. What are the advantages and disadvantages of sensitivity training to bring about racial awareness?

part 3
Women's Equality

We are here to claim our rights as women, not only to be free, but to fight for freedom. It is our privilege, as well as our pride and our joy, to take some part in this militant movement, which, as we believe, means the regeneration of all humanity. Nothing but contempt is due to those people who ask us to submit to unmerited oppression. We shall not do it!

— *Christabel Pankhurst*[1]

HISTORICAL VIEW OF WOMEN'S STRUGGLE

Some 52 percent of the United States population is female, and growing numbers of this majority are making startling demands upon all aspects of the social order. This new upsurge of the feminist movement is a continuation of an American effort that began in the early nineteenth century. The Grimké sisters — Sarah and Angelina — who can probably be credited with beginning the feminist movement, were the two daughters of a slave-holding South Carolina family. They set out initially only to speak against slavery, an institution they abhorred. In their time, the 1830s, however, this proved inordinately difficult as women did not engage publicly in reform movements nor address crowds comprised of both men and women.

Later, when the Grimké sisters left home to continue their fight against the slave system, they discovered how fettered they were as females in a male-dominated world. Martin Luther had made a pronouncement that characterized many prevailing attitudes of the male world they confronted. "Men have broad and large chests, and small narrow hips, and are more understanding than women, who have but small and narrow chests, and broad hips, to the end they should remain at home, sit still, keep house, and bear and bring up children."[2] In a letter to Mary S. Parker, president of the Boston Female Anti-Slavery Society, Sarah Grimké answered such attitudes in religious coin:

> *Here then I plant myself. God created us equal; — He created us free agents; — He is our Lawgiver, our King and our Judge, and to Him alone is woman bound to be in subjection, and to Him alone is she accountable for the use of those talents with which her Heavenly Father has entrusted her.*[3]

From the indignant reaction of the Grimké sisters, the seed of the feminist movement sprouted and expanded to what now can only be described as astounding proportions.

As the abolition movement progressed, other able women were attracted to the cause, but they quickly discovered that blacks were not alone in lacking the full rights of citizens. The editors of *The History of Woman Suffrage* summarized the problems early female fighters for equality faced. The most liberal of men, they maintained, found it impossible "to understand what liberty means for women. Those who eloquently advocate equality for a southern plantation cannot tolerate it at their own fireside."[4]

Not all opposition, however, was voiced by the male segment of society. Catherine A. Beecher, an educator, admonished her crusading sisters: "Heaven has appointed to one sex the superior and to the other the subordinate station. . . . It is therefore as much for the dignity as it is for the interest of females in all respects to conform to the duties of this relation."[5] Despite all opposition, an effort was instituted to organize against female inequalities. At a convention called in 1848 by Lucretia Mott and Elizabeth Cady Stanton at Seneca Falls, New York, American feminism as an organized force was born. At this meeting, a resolution was adopted demanding the right to vote, equal education, equality under divorce laws, and the right to preach, teach, write, and earn a living.

Shortly after this convention, Susan B. Anthony joined the movement and devoted the remaining fifty years of her life to the vigorous pursuit of the cause of women's rights. She was reared a Quaker, and initially her interest was in temperance, but her liberal Quaker friends realigned her efforts first toward the antislavery movement and subsequently toward women's rights movements. Her work included petitioning the New York legislature for women's property rights and the vote, and, in 1860, winning for the women of New York control over their wages and the guardianship of their children. She was instrumental in the demands made by teachers for higher wages and a voice in teachers' conventions. During the Civil War, she organized the Women's National Loyal League to press for the emancipation of blacks. In the years 1868 to 1870, Miss Anthony published *The Revolution,* a liberal weekly, in New York. The motto that graced its masthead was: "The true republic — men, their rights and nothing more; women — their rights and nothing less."

When the Fourteenth and Fifteenth Amendments to the Constitution were proposed to extend civil rights and the vote to male blacks, she demanded that this provision also be extended to women. When she failed to achieve this goal, she claimed and exercised her right to vote under these amendments as a citizen and as a person. By so doing, it was her intent to clarify the legal status of women by forcing a court decision. Instead, when she voted illegally, she was arrested, tried, and fined. When she refused to pay the fine, however, it was allowed to lapse. From then on, Ms. Anthony campaigned for a federal woman suffrage amendment through an organization she founded in 1869, the National Woman Suffrage Association. Later she served as president of the National American Woman Suffrage Association from 1892 until 1900, at which time she retired at the age of eighty.

Susan B. Anthony, in collaboration with three other advocates of female equality, compiled and edited the four-volume work, *The History of Woman Suffrage,* during the years 1881 to 1902. She organized the International Council of Women in 1888, and in 1900, in Berlin, Germany, with the help of Carrie Chapman Catt, she formed the International Woman Suffrage Alliance. When she died in 1906, woman's suffrage had been established only in Wyoming, Utah, Colorado, and Idaho, also in New Zealand, and Australia, but she had blazed the trail that led to the adoption in 1920 of the Nineteenth Amendment granting woman suffrage. Carrie Chapman Catt assessed the price of this culminating victory for female equality:

> *To get the word "male" in effect out of the Constitution cost the women of the country 52 years of pauseless campaign . . . 56 campaigns of referenda to male voters; 480 campaigns to get legislatures to submit suffrage amendments to votes; 47 campaigns to get state constitutional conventions to write woman suffrage into state constitutions; 277 campaigns to get state party conventions to . . .[adopt] successive congresses.*[6]

When the vote finally did come, women were accorded legal standing and civil dignity, but the miraculous transformation envisioned by suffragettes failed to materialize. It was some time before women used the ballot effectively to achieve in practice what they had advanced so tortuously from principle to law.

PSYCHOLOGICAL IMPLICATIONS

John Stuart Mill regarded women as a subject class. However, he recognized that the state of female bondage in at least one respect was a refinement over that of the black slave; each man wants his woman to be "not a forced slave but a willing one, not a slave, merely, but a favorite." In *The Subjection of Women,* he emphasized that subtle and pervasive social conditioning is the means by which women are prepared to accede to roles as the servants of men.[7]

Sigmund Freud was critical of Mill's study of women, believing that it gave insufficient consideration to what Freud construed to be inborn temperamental differences between the sexes. Sigmund Freud said, "Despite my 30 years in research into the feminine soul, I have not yet been able to answer . . . the great question that has never been answered: What does a woman want?"[8]

One of the answers that those active in contemporary women's liberation movements would give is cessation of the stereotypes of female status — the casual acceptance that women are less than human. Simone de Beauvoir, one of the intellectual stalwarts of the women's equality movement, wrote in *The Second Sex* that society, being codified by man, decrees that the female is inferior. Betty Friedan in her 1963 best-seller, *The Feminine Mystique,* documented the problems of middle-class housewives and observed that the American society seems bent upon persuading the housewife that all that is requisite for the fulfillment of those of her gender can be found in the cleaning, cooking, and child-rearing climate of the home. John Stuart Mill observed:

> *When we consider the positive evil caused to the disqualified half of the human race by their disqualification — first in the loss of the most inspiring and elevating kind of personal enjoyment, and next in the weariness, disappointment, and profound dissatisfaction with life, which are so often the substitute for it; one feels that among all the lessons which men require for carrying the struggle against the inevitable imperfections of their lot on earth, there is no lesson which they more need, than not to add to the evils which nature inflicts, by their jealous and prejudiced restrictions on one another.*[9]

Many of the roots of the contemporary women's rights movements stem from the prejudices and frustrations women encounter regularly when they attempt to leave their "designated" role and enter the world of work, research, or study. Psychologically and professionally, many find that established legal principles are not operative in daily practice in the outside world of trade and service. Perhaps the most damaging of all the handicaps a woman faces when she enters that world is the general assumption that a man by his very nature is capable of more than she is, and in every respect. The subtle psychological implications of this are reflected in early toys and unwittingly absorbed in childhood. Toys are constructed to imply that boys are activistic and will grow up to create and produce. Girl's toys, on the other hand, cater to a more passive nature and point toward a feminine role meant to nourish and consume.

Further psychological reinforcement for submissiveness is added when females are admonished that having an intellect may in fact be a hazard that will discourage proposals of marriage — the only worthy, ultimate goal for women. The effectiveness of this conditioning would tend to be supported by the fact that girls outperform boys in academic work *until* their late teens, when culturally established goals of marriage assume paramount importance, and a degree of reversal occurs.

Despite a fairly pervasive effort over a long period of time to predetermine the role and direction women should take, the results achieved, both past and present, have been far from uniform. Some women are very interested in female equality, some are apathetic, and some stand firmly against it. Kathleen M. Snow concluded, "Some women confess that their pro-liberation thinking is at odds with their anti-liberation feelings; they have been intellectually persuaded but their hearts belong to the old order."[10]

Pat Crigler, a psychologist at Northwestern University in Chicago, became very interested in why the active feminist role attracts some women and repels others. Finding no satisfactory answers at hand, she set about querying 750 women in Atlanta and Chicago. Two groups were selected for this study: the liberal National Organization for Women (NOW), and a more conservative organization, the League of Women Voters. Psychologist Crigler reported: "If I were going to say what has the most bearing on why a woman becomes a feminist, my answer would have to be that when there is a girl in the house who is for women's rights, it is the influence of her father."[11]

This study further indicated that the more education the father has, the more likelihood there is that the daughter will be a feminist. Still another significant factor determined by this study is that the higher the educational level in the home, the more likely girls are to believe in equal rights. An only child, it was determined, has a natural impetus toward a role beyond that of the traditional woman.

> *If you are the only child, you have to try to live up to the expectations of both parents. This means you learn both to sew and to fix the car. You are the confidante of both parents, and you see both sides of the fence. So you pick out the logical position for women, rather than the traditional.*[12]

Nonfeminists, the study showed, were likely to be women busy with large families. The more children a woman has, particularly if she does not work outside the home, the more likely it is that she is not going to be in favor of equal rights. From this study, the author concluded that there is considerable difference in perspective between those advocating equal rights and those active in women's liberation. There is a broader philosophical base for current trends than some of the most vocal spokespersons in both the pro and the con fringe areas of women's-equality movements would have us believe. Crigler concluded that the term "feminism":

> . . . *is a misnomer for what we are seeing. It is not women's lib, it is not the battle of the sexes. These are very poor descriptions of an interest in equal rights for all — of everyone striving to be accepted as a person and not be stereotyped.*[13]

PHILOSOPHY, FEALTY, FRUITION, AND FAILURES

The rebellion against the stereotypes of female status and against the casual assumption that women are inferior human beings was the impetus for the formation of organizations to dispel such myths. The role of women in the labor and military forces in the World War II era and after that time made attitudes and acts of inequity for women of more than theoretical concern to those who were subjected to them. The increased number of women who did enter the work force was an important element in the institution and growth of the feminist movement of the 1960s and 1970s.

A dramatic change occurred in the profile of the female labor force from a preponderance of unmarried young women to a majority of older women. By 1968, almost 40 percent of all working women were at least thirty-five years old. Whereas the younger women, who were expecting to work only until marriage or their first pregnancy, were not as avid about raising the issues of limits placed upon their earnings and aspirations, older women expressed a growing resentment toward economic and career inequities. This resentment was justified in fact:

> *The Women's Bureau, a division of the Department of Labor, has found that, regardless of the equal pay provisions of Title VII of the 1964 Civil Rights Act, there is a double standard of wages for full-time, year-round workers; without exception, the national average of annual wages is much higher for men than for women.*[14]

Although the practice of bra-burning may be passé, the women's liberation movement has directed the consciousness of vast numbers of people to the issue of sexual equality. Their underlying goal — to recondition the American people to accept sex equality as a norm of social and personal behavior — has been partly achieved and continues to make progress. Caroline Bird, author of *Born Female,* states that "after a long period of celebrating the difference between men and women, we are heading into an androgynous world in which the most important thing about a person will no longer be his or her sex."[15]

Domestic frustrations and the hostility of women who abandoned the hurdles of housewifery only to encounter barriers to equal pay and professional advancement in the world of work, plus the failure of earnest efforts of organizations such as the Kennedy Commission on the Status of Women to redress their grievances — all contributed to the formation in 1966 of the National Organization of Women (NOW). The problem of female inequality had lain buried and unspoken for many years in the minds of a large segment of American women.

> *It was a strange stirring, a sense of dissatisfaction, a yearning that women suffered in the middle of the twentieth century in the United States. Each suburban wife struggled with it alone. As she made the beds, shopped for groceries, matched slipcover material, ate peanut butter sandwiches with her children, chauffeured Cub Scouts and Brownies, lay beside her husband at night — she was afraid to ask even of herself the silent question — "Is this all?"*[16]

Some investigators of the National Organization of Women say that the effort is as much a state of mind as it is a movement. However, NOW president Wilma Scott Heide described the women's liberation movement as "the most profound universal behavioral revolution the world has ever known."[17]

The NOW version of the women's equality movement has been likened to the civil rights movement for black liberation and is said by many to parallel it. It is true that women's protests followed black protests of the 1960s:

> *Antislavery movements preceded the first coherent women's rights movement, black male suffrage, woman's suffrage, the civil rights movement, the new feminism. White women it appears have organized not after working with blacks, but after they have worked in behalf of them. That blacks and women should have a common enemy, white men and their culture, without making common cause is grievous, perhaps. They even have more in common than an enemy. In America they share the unhappy lot of being cast together as lesser beings. It is hardly coincidence that the most aggressively racist regions are those most rigidly insistent upon keeping women in their place, even if that place is that of ornament, toy, or statue.*[18]

Among early states that refused to ratify the Nineteenth Amendment, nine out of ten were southern.

The contemporary women's liberation movement is divided roughly into three major camps, although these may, at times, interlap and are in a continuous state of flux. Roughly, the divisions are the conservative feminists, the politicos, and the radical feminists. The conservative feminists are characterized by members of NOW and a myriad of splinter organizations that hold similar philosophies. In the beginning, NOW was composed of older professionals and career women who had achieved some modicum of success. It concentrated on the surface symptoms of sexism — legal inequities, employment discrimination, and so on. Its stress was placed on equality within the given system rather than radical departure from mores such as family values. Recently, however, with the influx of younger members, the philosophy of mild reform has been abandoned for less cautious issues.

The politicos of the contemporary women's movement are those women whose primary loyalty takes priority over the women's liberation movement proper. They view inequality of women as merely a small piece of the larger whole that recognizes the entire world as divided into two camps, the powerful and the powerless.

The radical feminists are often drawn from the ranks of disillusioned moderate feminists from NOW or disillusioned leftists from the liberation movement *plus* an unaligned segment that considers feminist issues not only as women's first priority but as central to any larger revolutionary effort. Shulamith Firestone stated:

> *If any revolutionary movement can succeed at establishing an egalitarian structure, radical feminism will. To question the basic relations between the sexes and between parents and children is to take the psychological pattern of dominance-submission to its very roots. Through examining politically this psychology, feminism will be the first movement ever to deal in a material way with the problem* [*of women's inequality*].[19]

On November 24, 1969, a group of more than 500 women, representing numerous women's lib groups, attended a congress to unite women in New York. The delegates agreed on the following demands:

> *1. Nationwide, free 24-hour child-care centers for all, staffed equally by men and women, with immediate income tax deductions for child-care expenses until they are in operation.*
>
> *2. Opening of all courses of study to boys and girls without pressure to elect on the basis of sex.*
>
> *3. Programs of women's study similar to black studies in all colleges.*
>
> *4. Flexible working hours for both men and women and part-time employment for women who want it.*
>
> *5. Investigation of the percentage of women hired in each job category of big companies.*
>
> *6. Abolition of the presumption that sex roles are biologically determined. "Children should be given human models to emulate, not just male and female models."*
>
> *7. Penalties for violation of discrimination on the basis of sex in employment prohibited under Title VII of the Civil Rights Act of 1964.*
>
> *8. Equal Rights Amendment to the Constitution of the United States striking down all laws that classify citizens on the basis of sex.*

9. Legal recognition of the right of every woman to determine whether she shall bear a child by the repeal of all laws against abortion, birth control, and sterilization; and free birth control to all women who want it.

10. Protest against the "generally derogatory image of women presented by the media, and specifically the misrepresentation of the movement for women's liberation to the women of America"[20]

The proposed Twenty-seventh Amendment this congress referred to includes in Section 1, "that equality of rights under the law shall not be denied or abridged by the United States or by any state on account of sex;" in Section 2, "that Congress shall have the power to enforce by appropriate legislation, the provisions of this article;" and in Section 3, "that this amendment shall take effect two years after the date of ratification."

The two foregoing documents specifically and generally characterize the women's demands for equality. Already feminist lobbying and litigation have destroyed many barriers to sexual equality, particularly in the field of employment. The academic world, long a bastion of male superiority, has been forced to take into account feminist pressure for equality. Female visibility in politics was more evident at the 1972 presidential nominating conventions, and the impact of female leaders such as Shirley Chisholm, black congresswoman from Brooklyn, is significant. The number of women appointed and elected to offices has been noteworthy, as has been the number of lawsuits that have succeeded in overthrowing sex-discriminatory practices.

The influence on print media has also been notable. There has been an unprecedented flow of books on the subject of women's equality as well as a proliferating list of periodicals, such as *Off Our Backs, Everywoman, Ain't I a Woman?, Up from Under, Tooth and Nail, No More Fun and Games,* and *Ms.* Even *Esquire,* a man's magazine that once banned female authors, began a woman's column and devoted its July, 1973, issue to women.[21] These achievements are evidence that one of the primary goals of feminism, which is to wake women up, is succeeding. The persistence of the movement, despite its divisions, is paying off. A sympathetic male observer, William Henry Chafe, history professor at Duke University, commented that:

> *Important shifts in behavior have taken place and the changes bear directly on some of the root causes of sexual inequality, the definition of male and female spheres, the role models we provide our children, the permissible horizons available to men and women.*[22]

According to Simone de Beauvoir, these changes in behavior may not be as reprehensible as some critics have anticipated: "To emancipate woman is to refuse to confine her to the relations she bears to man, not to deny them to her; let her have her independent existence, and she will continue nonetheless to exist for him also; mutually recognizing each other as subject, each will yet remain for the other an *other.*"[23]

Although the movement for women's equality has recruited many devotees and achieved almost phenomenal success, it has failed in many instances to persuade large numbers with its message. These people question the logic and effects that may ultimately emerge from its success. Critics argue that the basic feminist premise — that there are no profound physical or emotional differences between the sexes — does not correspond to the reality of daily relationships. George Gilder has warned that the male social role requires an economic advantage: "A woman with more money than the men around her tends to demoralize them. . . . She weakens their connections within the community and promotes a reliance upon other anti-social ways of confirming their masculinity."[24]

Opponents of the Equal Rights Amendment feel that existing federal statutes adequately guarantee equal rights in job and wage areas. They voice concern about such matters as child and wife support. They anticipate legal complications about such conventional and routine matters as separate public toilet facilities and separate prisons. Persistently, however, women are forcing the social order to live up to the promises of equal opportunity and protection guaranteed by the Constitution and the law.

To those whose work it is to further the cause of women's equality, it will be necessary to view marriage as the changing institution it has become; to evaluate the status of women *both* idealistically and realistically; to examine the implication of child-care demands; to deal with the concerns of rape and legalized prostitution, coeducation, and child adoption. The new personal and economic liberation of women entails these concerns. Changes in women's equality are coming and will continue to come. The relationships and potentials that emerge will depend in large measure on the wisdom of those who midwife the current female travail:

> *In the radical feminist view, the new feminism is not just the revival of a serious political movement for social equality. It is the second wave of the most important revolution in history. Its aim: overthrow of the oldest, most rigid caste/class system in existence, the class system based on sex — a system consolidated over thousands of years, lending the archetypal male and female roles an undeserved legitimacy and seeming permanence. In this perspective, the pioneer Western feminist movement was only the first onslaught, the fifty year ridicule that followed it only a first counter-offensive — the dawn of a long struggle to break free from the oppressive power structures set up by nature and reinforced by man.*[25]

As of July 1, 1974, the Equal Rights Amendment (ERA), also known as the Women's Rights Amendment, had been ratified by thirty-three states, although two of them, Nebraska and Tennessee, slid backward and rescinded their legislative approval. That leaves five or seven to go, depending on how the legal battles in those two states are resolved. Women everywhere, service women and civilians, look eagerly toward the near future for the passage of this amendment, which would prohibit the enactment of any laws that would discriminate on the basis of sex.

WOMEN IN THE MILITARY

An overview of women's equality would be incomplete without a consideration of the military, which for centuries has been considered an exclusively male domain. In the United States, women have been in the military officially for only about thirty years. In comparison with men, however, there is a decided paucity of information and data about female military personnel and problems. Since at their peak of participation in the military (1945), women represented only 2 percent of the total forces, there is some reason for the neglect or omission of women from past military analyses. With new emphasis being placed on female recruitment for the military, however, decision-makers find themselves in the position of having to make judgments on the basis of incomplete or nonexistent data. Concerted efforts are being made to remedy this problem.

An examination of the status of women in their relationship with the armed services finds them categorized in one of two ways: either "in" the armed forces, or "with" the armed forces: "The word *in* when referring to women is used in reference to their being integrated into the service . . . the word *with* when referring to women is used whenever women have been in an auxiliary status. Auxiliary status means that women are not granted the rights and benefits of those on full military status."[26]

The first women to serve with the armed forces in any numbers were nurses under civilian contract during the Spanish-American War. In recognition of their service, the Army Nurse Corps was established by Congress in 1901, and the Navy Nurse Corps in 1908.

Aside from those in the field of health, the first women in the United States military served in the Navy during World War I and were called "yeomen." These women gained access to the service through a legal loophole. The law governing the Army specified the enlistment of male persons, but the law pertaining to the Navy stipulated only the enlistment of *citizens*. Josephus Daniels, then secretary of the navy, was enthusiastic about getting women into the service, and when they sought enlistment, he summoned his legal aides and asked them if there was any reason why women should not be in the Navy. Finding there was no legal basis on which to exclude women, Secretary Daniels opened the Navy to women under conditions of equal opportunity. To appreciate the full implication of this decision, it should be remembered that his pronouncement was made in 1918 and women were not given the vote until 1920. In all, some 13,000 women served in the Navy and Marines during World War I. In the unequal climate of that time, most were assigned clerical positions. Although there were numerous requests from Army commanders, all efforts were fruitless to modify the law so that women could be enlisted in the Army.

During the interwar period, the position of Director of Women's Regulations was established by the Army to coordinate efforts to plan for a women's corps. This position was terminated by General Douglas MacArthur, who stated that he considered the director's duties to be of no value to the Army. The plans which had been developed to set up a women's corps to be *in* the Army were shelved and turned up again only after the Women's Army Auxiliary Corps (WAAC) had been in operation for six months.

During this same interim, the loophole that had allowed the Navy to enlist women was closed by the simple insertion of the word "male" before the word "citizen" in the description of those eligible for enlistment in the Naval Reserve Act of 1925. The modification was retained when the legislation to establish a women's corps was debated — whether or not such a female contingent should be *in* the Army or serve *with* the Army as an auxiliary. After prolonged Congressional debate, the bill establishing the WAAC finally passed in May, 1942, almost a year after its introduction. Full military status was granted to women in 1943 for the duration of the war plus six months — and the name of the organization was changed to the Women's Army Corps (WAC). A bill for a women's corps to be *in* the Navy shuffled through Congress in the face of opposition, but was eventually passed in the early summer of 1942. These female Navy forces were called Women Accepted for Volunteer Emergency Service (WAVES). There was recruitment difficulty in both corps partly attributable to the adverse attitude of men in the service toward the women's corps. Also there was initial discontent on the part of the women at misplacement and disillusionment with the types of jobs assigned, which were mostly menial.

Despite adverse public and private attitudes regarding women in the armed forces, at the apex of World War II, over 265,000 were in service and evidence abounds that they did play an important role within the military establishment:

> *In recognition of this service and for the continuing contribution women could make in the Military Establishment, in 1948, President Truman signed Public law 625 authorizing the establishment and appointment of women in the Regular Army, Navy, Air Force, and Marine Corps.*[27]

A review of the past indicates that with few exceptions the common attitude toward the women's corps has been consistently one of acceptance, but only as a last resort. It has been said that history repeats itself, and it remains to be seen how much the services have learned from their past experience and what measure of equality there is to be for women in the new all-volunteer force. The armed forces have never been able to fill their needs for manpower without conscription, but availability of the draft to meet manpower needs ended with the expiration of the Selective Service Law in July, 1973. Some inequalities requiring resolution are that women must meet higher standards for admission to the armed forces than men and that women must be high school graduates or have a general education development (GED) equivalent to high school completion.

Women are accepted into the service only from Mental Groups I, II, and upper III, whereas men are accepted in Mental Groups I to IV. Mental groups I, II, and III are the highest categories of the military intelligence test performers. Additionally, the battery tests for enlistment which women take are academically oriented, whereas the battery tests which men take are aptitude oriented. Past records show servicewomen have worked primarily in only two of the eight military occupation groups — sixty-seven percent in administration and twenty-two percent in health. Recently, however, the range of occupations and training open to women in the services has been significantly increased, and despite some continued inequities, there is a new spirit for full equality for women in the armed forces.

Several rules concerning the marriage, pregnancy, and "moral character" of women in the armed forces or of prospective enlistees have been attacked by groups outside the military, including the American Civil Liberties Union, on the ground that similar restrictions are not applied to men.[28] In response to exposure of this female inequity, the Army has canceled several restrictions on the marriage and pregnancy status of women who want to join the WACs. Married women will be considered for enlistment on the same basis as single women. It is no longer necessary for women who have had an unwed pregnancy to apply for a waiver as a requisite for enlistment. The requirement that women submit three character references on enlistment applications has also been dropped.

Section 6015, Title 10, of the United States Code states: "Women may not be assigned to duty in aircraft that are engaged in combat missions, nor may they be assigned to duty on vessels of the Navy other than hospital ships or transports."[29] A strong advocate of women's equality, Admiral Elmo Zumwalt, Jr., former Chief of Naval Operations, issued Z-Gram 116, which some call the Navy's own Equal Rights Amendment, so that the Navy upon passage of the amendment could assign women to all jobs that sailors now perform. Also, when the U.S.S. *Sanctuary* was given extensive overhaul and conversion, in December, 1973, two women officers and sixty enlisted women made history by becoming part of the five hundred-person crew.

Besides opening the possibility of assigning women to ships, the Navy's new policies toward its women members have opened other previously closed areas in which female enlistees will now be able to operate. One of these is the Reserve Officer Training Corps (ROTC) programs on many of the nation's college campuses. In the fall of 1973, seventeen coeds at four universities became the first women to enroll in the program. Officials of the Navy are also making major efforts to open up even more programs previously underused by women. The current emphasis is to enroll more enlisted women in the Navy Enlisted Dietetic Education Program (NEDEP), and the Navy Enlisted Nursing Education Program (NENEP). These three college programs can help persons acquire a degree, a commission, and a wider choice of career fields.

In the past, women have been allowed to participate in some department courses, but without the full ROTC credit. This inequality was remedied June 4, 1973, when it was reported that the University of Vermont's Department of Military Studies would permit freshmen and sophomore women to join the Reserve Officer Training Corps program with the full status of men students. After graduation they will be eligible for a commission in the Army, National Guard, or Army Reserve.[30]

Another evidence of inequality for women in the military has been the prejudicial treatment of female members of the armed forces in the application of dependency criteria. A regulation change was approved, effective May 14, 1973, which authorized female members "to claim their civilian husbands as dependents for the basic allowance for quarters without regard to proof of in-fact dependency." Further, Department of Defense officials have declared that they support past legislative proposals which would provide equality for female members in the application of dependency criteria, in accordance with the Department of Defense Human Goals Program.

Also in 1973, the Army opened more of its specialties (90%) to women:

> *Army* WAC's *have become military police, truck drivers, heavy-equipment operators, and are in training to become parachutists and helicopter pilots.* WAVES *are eligible for 66 of the Navy's 88 job ratings. They are serving as deck hands, electronic specialists, band members, and chaplains.*[31]

The Air Force has opened about 98 percent of its skill fields to women and the Marine Corps has opened 57 percent. It is also possible now for a woman in uniform to stay in the service after the advent of motherhood. The rule banning mothers was amended in April, 1971, and now there are almost 1,000 mothers in uniform.

These examples of past inequalities for women serving in the military and the large numbers of them that are being set aside exemplify the new attitude of those in command toward the importance female recruitment holds in the new, all-volunteer forces.

Women Officers in the United States Army: Liberated?

Major Sherian G. Cadoria, Executive Officer

WAC Branch, Office Personnel, Directorate Department of the Army, Washington, D.C.

The period of my Human Relations master's degree internship with the Department of the Army, Women's Army Corps Branch, was filled with many diversified activities that proved to be challenging and rewarding experiences.

In order to consolidate my experiences, I have focused this report on three areas: (1) the process of integrating women officers into the "male" branches of the Army; (2) the black woman officer; and (3) married women officers.

INTEGRATION OF WOMEN OFFICERS

On April 17 and 18, 1974, the Branch Selection Board was convened at the Department of the Army to select women officers for integration into the male branches of the Army. Each woman had previously submitted three choices of branches in which she had an interest. The reaction of the twelve men sitting on the board was one of astonishment — astonishment because this was the first time they actually saw how women had been "held back." The women's records overall were not competitive with their male counterparts. Women in the past had been restricted to clerical duties or serving as billeting officers, nursery officers; however, as the woman advisor to the board, I found it for women to display the intelligence and intellect they awaited the opportunity to display. Nonetheless, it was necessary for each branch to accept its share of the women officers, to include high-quality and low-quality applicants as shown by their files.

Persuasion was my overall method of getting the men to accept these officers; however, as the woman advisor to the board, I found it necessary to display forcefulness and aggressiveness heretofore unnecessary to employ. Leaving the board room on the evening of April 18, the president of the board stated, "World War IV has been waged here, gentlemen, and the victor is Major Cadoria."

Yet victory was still to be attained. This was just the first step in a long climb to equality. On July 1, 1974, I assumed a new position, that of "women officer monitor." This position was created to insure that our 1,300 women officers (a figure that is increasing daily) would be utilized according to their skills and training so that they would be competitive with their male counterparts. This position involved setting up an automatic data processing system to monitor all assignments. It also required worldwide travel and congressional and public appearances. I feel that my selection to fill this position represented another achievement for black women officers.

MARRIED WOMEN OFFICERS

When one stops to think about the state to which our society has evolved, there is no reason in theory, sociology, or equity that women should not have every opportunity that men have. Any man or woman should be permitted to serve his or her country in any capacity that he or she, as an individual, views as appropriate.

When the Women's Army Corps was established in May, 1942, the Army immediately established regulations prohibiting women from serving on active duty with children or dependents under age eighteen. Married women on active duty could not claim their husbands as dependents or obtain privileges for their husbands (such as use of commissaries, post exchanges, or medical facilities); women could not draw BAQ (cost of living allowance or housing allowance) for their spouses. Therefore, the number of women who married and remained in the Army could be counted on both hands.

Today, doors formerly closed to women are being reopened; discriminatory policies which barred married women from the Army and forced automatic discharge for women who became pregnant were eliminated in 1971. Another stride was made in 1972, when Congress passed a bill allowing women in the Army to claim their husbands as dependents; in 1973 it was ruled discriminatory to withhold basic allowance for quarters for female married officers. The result of this action was a rush comparable to the gold rush (only this time it was the wedding-band rush). Today approximately 15 percent of the women in the Army are married.

The impact of these changes has resulted in certain problems not encountered by the male Army. How do they affect the management of women officers? Are these problems a result of discrimination or prejudice?

Let me now point out that although I have noted some of the changes made to end discrimination against women, and especially married women, there are some practices which have not been eliminated, such as the case of a pregnant officer who has to request permission to remain on active duty; a single officer who has to request permission to adopt a child; and rules about nonfraternization between enlisted soldiers and officers. These practices present many problems that tax the ingenuity of the personnel-management officer.

When one speaks of human relations, many ideas and concepts come to mind. However, to me, human relations as it involves men and women, single or married, is more than just a term. It is active understanding, compassion, objectivity, and plain "humankindness" when dealing with other individuals. Some of the problems faced by married women are depicted in a series of what I refer to as "case studies" (see Case Studies 1 to 5 at the end of this report). These are but a few of the many cases handled daily by my office. Of note, the number of female officers marrying enlisted soldiers is increasing at such a rapid rate, and causing such problems for both parties involved, that all branches of the Army are becoming very concerned.

Perhaps you will ask, why not abolish the nonfraternization policy (forbidding an officer and an enlisted person from participating in *planned* social functions or maintaining a relationship beyond what is required to accomplish the official mission)? The courts in Atlanta, Georgia, have upheld the legality of the Army's policy of nonfraternization. Officer and enlisted couples are therefore confronted with restrictions of assignments, limited on-post housing, and very restricted social activities, since enlisted personnel are not allowed in officers' clubs or permitted to attend their social functions. Further, a woman officer married to an enlisted soldier cannot command units at training centers or perform recruiting duty. These are but a few limitations placed on career and professional development. It is clear that a problem exists; the dimensions of it are less clear. Research underway by the Department of the Army should provide some answers and perhaps result in an alignment of policies consistent with today's social changes. For now, however, sexism in the Army is not a byword; it is a reality.

I feel that comprehensive information programs should be initiated so that the young woman anticipating enlistment in the Army is aware of the difficulties she may encounter; that more aggressive steps should be undertaken by the Department of the Army to study current, outmoded policies; that the women who elect marriage and the Army recognize that there will be separations from spouses just as the male officer must sometimes be separated from his family.

BLACK WOMEN IN THE ARMY

From 1942 to 1954, hundreds of black women served in the military forces. Their jobs were menial; they were laundresses, cooks, and the "mop pushers." In 1961, at the home of the Women's Army Corps in Anniston, Alabama, some seven years after President Harry S. Truman desegregated the Army, black women officers were still being treated as inferior. For example: black officers were not allowed to command a troop unit at the training center or fill the prestigious position of platoon officer or executive officer of the student officer company. Command positions were filled worldwide by white officers, while blacks were assigned to positions of administrative or clerical work. Few blacks were allowed to attend Command and General Staff College (the quota for any woman was minimal) or selected for the civil schooling program (master's or doctor's degrees). A vacuum was thereby created so that in 1969, when Congress opened up all ranks for military personnel, male or female, there were no black women officers in the position to be promoted to general officer. Of note, no black woman officer has ever commanded the U.S. Army WAC Center and School, or been selected to attend the Army War College (considered a prerequisite for attaining the rank of general officer).

As of this writing, there are eighty-one black women officers in the U.S. Army: one colonel, who holds the position of Deputy Commander, U.S. WAC Center and School; five lieutenant colonels, one of whom is the first black woman officer to command a battalion at the U.S. WAC Center and School; twelve majors, three of whom hold positions as *first*, that is, commander of the Student Officer Company, commander of the Reception and Processing Station, and executive officer of the Women's Army Corps Branch; eleven captains and fifty-two lieutenants, one of whom holds the position as the first black women aide-de-camp to a general officer. Strides have been made, but there is room for far more. It is my contention that special action should be taken to insure that the black woman officer has a chance to push forward. My efforts have been channeled in that direction.

Questioning 44 of the 81 black women officers about discriminatory practices, I found that most felt that they experienced more discrimination because they were women rather than because they were black. This is related to what may be called a female bias, an image that is prevalent throughout the Army. Working under supervisors of different racial backgrounds did not appear to pose significant problems.

There is much to be done before equality comes to women, and, when they are black, the efforts to achieve equality are doubled.

In conclusion, I can only say that working at this level, striving to achieve an equitable share of what the Army has to offer and insuring fair and just practices and policies for the women officers has been one of the most rewarding experiences for me to date. The following case studies highlight the complexity of my job.

CASE STUDY 1

Name: 1LT A; Race: Black; Assignment: Commander, Fort Sam Houston, Texas; Spouse: 1LT B.

Background: Lieutenant A entered the Army in 1971 as a single, second lieutenant. She was assigned to Fort Sam Houston, Texas, following completion of the WAC Officer Basic Course. On July 1, 1973, because of her dynamic leadership and demonstrated potential, she was assigned as commander of a battalion.

In 1972, Lieutenant A married B, a military intelligence officer, who was enroute to Korea (a short, unaccompanied tour) at the time of marriage. On August 6, 1973, Captain X, Military Intelligence Branch, contracted WAC Branch regarding the reassignment of LT A to Fort Bragg to be with her husband, who was being reassigned there from Korea in January, 1974. At that time, an affirmative response was given.

On August 20, 1973, LT A contacted Major Cadoria, her assignment officer, to inform her of her husband's impending return to the United States and her desire to be detailed to the Quartermaster Corps. Her file was screened and she was accepted for detail and subsequently assigned to Fort Bragg. On September 20, 1973, however, the Army went into an expansion of the female enlisted corps, necessitating a hold on individuals serving in command and troop positions in addition to increasing the number of troop units. On same date, LT A was notified that due to the approved expansion of the WAC, there was no suitable replacement available to assume command of her unit November, 1973, but every effort would be made to reassign her in the January/February time frame.

On October 29, 1973, LT A's file was submitted to Quartermaster Branch for detail and orders to be published assigning her to Fort Bragg with Temporary Duty enroute to the Quartermaster School to attend the Supply Officer Management Course to qualify her for detailed assignment. LT A was informed of this action and stated that she was pleased with the action taken. She stated that she also understood that command assignments have top priority and that there were very few black officers in positions which the enlisted troops felt were highly visible. She said she was happy to "help out" at a time when every qualified officer would be called upon for troop duty. (There are only 81 black female officers in the WAC, whose total strength is 1,300 officers.) Without visibility in certain positions such as command, the WACs face human relations problems which could be magnified by the women coming into the present environment as well as those already a part of the environment.

Subsequent to this discussion, a back-channel message was received from a general officer which inferred that neither the WAC Branch nor Military Intelligence Branch was showing any compassion to these two officers, and that the Army was going to lose two highly qualified black officers. Although communication had been maintained between both LT A and her husband, LT B, by the assignment officers, they, A and B, alleged failure to receive any consideration. Communication by telephone was arranged between A and B and the three branches involved in this action. The assignment officers were later directed to reassign LT A in December, 1973, so that she could be with her husband immediately.

Action Taken

LT A was reassigned with a departure date from Fort Sam Houston, Texas, of December, 1973.

Effect of Action

1. No black women officers in positions of high visibility in a unit where both white and black personnel insist on black visibility.
2. Uprooting three officers to satisfy the whim of one person.
3. Retention of two black officers, one male and one female.
4. Resentment by single officers of special attention afforded married officers.
5. Another demonstration of the compassion and understanding of the Department of the Army.

CASE STUDY 2

Name: 1LT C; Race: Caucasian; Assignment: Commander, WAC Company, Fort Monroe, Virginia; Spouse: Civilian (occupation continuously changing).

Background: Lieutenant C entered the Army in 1970, as a single officer, and married in 1972. Her performance has been truly exceptional and she has a great future ahead of her in the Army. Assigned to Fort Monroe in 1971, she was allowed to remain there after her marriage in 1972. In July, 1973, she requested assignment to Fort Jackson, South Carolina, as her husband had applied and been accepted for Nursing School in that area. Orders were immediately published reflecting a January report date as her husband would start school in that time frame. (A single officer was designated to replace LT C.)

In mid-October, LT C informed me that her husband had changed his mind and did not want to go to school but was going to attempt to start a business of his own. She also stated that if she left the area, her marriage would in all probability end in divorce as her husband had issued an ultimatum to her. Personal contact was made with the husband, who absolutely refused to consider his wife's career.

LT C, on several occasions, reiterated to me that she wanted to remain in the Army and that perhaps at a later date, she could change her husband's attitude.

Action Taken

LT C was allowed to remain at Fort Monroe for an additional period of one year.

Effect of the Action

1. Resentment of the single officer scheduled to replace LT C at Fort Monroe and subsequently reassigned to Fort Jackson.

2. Congressional appeal by the single officer to prevent her assignment to Fort Jackson.

3. Telephone calls from both male and female officers protesting the reassignment of the single officer to Fort Jackson.

CASE STUDY 3

Name: 2LT D; Race: Caucasian; Assignment: Drug and Alcohol Control Officer, Fort Hamilton, New York; Spouse: Civilian (Student).

Background: Lieutenant D entered the Army in January, 1973. Following the Women Officers' Orientation Course, she received training as a Military Intelligence Officer at Fort Huachuca, Arizona. Married while in school, she was offered an assignment at Fort Dix, New Jersey, the nearest location to her husband with an open position in her military occupation specialty. She refused the assignment because she wanted to be with her husband in New York. Therefore, she was assigned to Fort Hamilton, where there were no positions for a military Intelligence Officer.

LT D became disenchanted with her job in a very short time. She requested reassignment to another job at Fort Hamilton to utilize her specialty skills. (No jobs in her skill area were available but the WAC Company needed a commander so this job was offered to her, *if* the post commander approved her reassignment.) Although LT D wanted the command position, the post commander refused to reassign her (once the Department of the Army assigns an individual, the installation or post commander controls assignments of his assets on his installation or post).

LT D was informed of the commander's decision. She then informed me that her husband had decided to return to Milwaukee, Wisconsin, following his December, 1973, graduation. I then requested, was granted authority, and reassigned LT D to Milwaukee as a recruiter so that she could be with her husband.

Effect of Action

1. Retention of an officer with great potential in the military.
2. Good public relations with the civilian community, especially that of the lieutenant's hometown.
3. Vacant position at Fort Hamilton, which requires some officer to work a dual job.

CASE STUDY 4

Name: Captain E; Race: Black; Assignment: Army Community Services Officer, Presidio of San Francisco, California; Spouse: Infantry Officer, U.S. Army (White).

Background: Captain E was married in 1972, after serving three years in the Women's Army Corps. In January, 1973, the Army needed Captain E for an assignment to the U.S. Women's Army Corps Training Center. She, of course, felt that the discrimination against her would be so great as to render her emotionally unqualified to perform her job. Captain E's husband is white and it was realized that in many areas, the couple would encounter such difficulties. It is also important to note that both training units or centers for women in the Army are located in the South. Fort McClellan where the first assignment was proposed is located in Anniston, Alabama.

Action Taken

1. Officer assigned to specifically requested location.
2. Single officer assigned to Fort McClellan as a replacement.

Effect of Action

1. Single officer who would have gone to California (considered one of the best assignments), had to accept a second tour at the training center.

2. Limited assignment areas for this officer or acceptance of separation from husband on some assignments.

3. If officer accepts only limited assignments, it will result in officer not being competitive with her contemporaries.

CASE STUDY 5

Name: 2LT F; Race: Caucasian; Assignment: Military Police, Fort Benning, Georgia; Spouse: Master Sergeant, U.S. Army.

Background: Lieutenant F entered the military in January, 1973, attended the Military Police (MP) School in April, 1973, and graduated number two in her class. She married an enlisted soldier during the course and requested to be reassigned with him. (The Army's nonfraternization policy between officers and enlisted has been upheld as legal by courts of law). Because of her standing and demonstrated performance, against the better judgment of the MP assignment officers, she was assigned to Fort Benning to be with her husband.

The Provost Marshal, a prejudiced white officer nearing retirement, refused to allow LT F to function as a military police officer. Attempts to change LT F's mind about remaining at Fort Benning were met with negative results. She insisted on remaining with her husband.

Action Taken

LT F remained at Fort Benning.

Effect of Action

1. LT F, a highly qualified military police officer, is not being properly utilized.

2. LT F received a very poor efficiency report from a very prejudiced senior officer. (This colonel was investigated for discriminatory practices against women).

3. LT F's career may be limited if she is unsuccessful in appealing the report mentioned in paragraph 2.

QUESTIONS FOR FURTHER CONSIDERATION

1. Do women have the military prowess for combat duty and should they be permitted to engage in it?

2. What will the psychological effects, if any, be on children if day-care centers and child communes become the order of the day?

3. Will passage of the Equal Rights Amendment polarize women into two camps politically and diminish their potential to act in this area as a block vote?

4. Will the sentiment for war be increased or diminished by the large number of female recruits presently seeking careers in the armed forces?

5. What steps can be taken to mitigate hostile attitudes on the part of male and female enlistees toward each other?

6. Should enlistees have common training accommodations from the beginning of service or only after initial training and skill proficiency have been completed?

7. If the Equal Rights Amendment is ratified, what modifications in public school education will be necessary to transform the letter to the spirit of female equality?

8. What changes have recently occurred in the military to eliminate some of the discriminatory practices discussed in this chapter?

part 4
Military Justice

Sergeant: You know what the trouble with peace is? It's nothing but disorder. And when do you get order? In a war. In a war, everyone registers, everyone's name's on a list. Their shoes are stacked, their corn's in the bag, you count it all up — cattle, men, et cetera — and you take it away!

— *Bertolt Brecht*[1]

HISTORICAL ORIGINS

As King Richard I of England set sail for the Holy Land in 1190, he foresaw the necessity for some statute to govern the behavior of his soldiers and sailors on the voyage and beyond: "Whoever shall slay a man on ship-board, he shall be bound to the dead man and thrown into the sea. If he shall slay him on land he shall be bound to the dead man and burried in the earth."[2] Richard's simple ordinance was the first directive promulgated on English soil to set down specific punishments for military offenses as distinct from civilian offenses. Little did Richard the Lion-Hearted know that his order would be the first in a long line of orders that would finally lead to the complex and controversial legal framework known as the Uniform Code of Military Justice of the United States.

Most legal historians agree that the ancestor of today's court-martial was the Court of Chivalry established by Edward I of England sometime before the close of the thirteenth century. The authority of this court, however, was limited since its jurisdiction "could not be exercised over soldiers 'within the realm in time of peace' if the central civilian courts were open and King's Writs could be issued."[3] Richard II, in 1385, issued Articles of War that contained twenty-four paragraphed items governing, among other things, the conduct of soldiers. Although a few English kings had enforced military law in peacetime before 1628, it was against common law practice, and Parliament in its Petition of Right addressed to Charles I in that year registered this as one of their major complaints against the Crown. After 1628, Parliament stated that court-martial jurisdiction could be enforced abroad only, and for only such serious offenses as mutiny, sedition, and desertion. It was further stipulated that any Articles of War authorized by the Parliament were enforceable only in time of war. English common law was a remarkably strong force, even in the realm of military discipline, for Charles V's Penal Code of 1532, the Articles of War of 1590, and the War Articles of Gustavus Adolphus of 1621 all treated the soldier as legally one apart.[4]

It was not until 1689 that Parliament, in its Mutiny Act, set down specific penalties that court-martials could enforce for such crimes as mutiny, sedition, and desertion.[5] The next significant piece of English legislation governing military offenses was the Articles of War of 1765, which were said to be a literal translation of ancient Roman Articles of

War. These articles, however, provided that "normally" civil authorities would try soldiers for capital crimes or for offenses against the person or property of civilians. Jurisdiction of military courts was restricted to offenses directly connected with military discipline, although power was granted to military courts to try such civil offenses as murder, rape, robbery, and counterfeiting *if* British civil courts were closed. John R. Thornock has pointed out, however, that "this caveat was tantamount to a requirement that total war, with attendant disruptions of civil services, would exist throughout the British Isles. Historically, this had, indeed, been a rare occurrence."[6]

The first articles of war promulgated for American soldiers were those adopted by the Provisional Congress of Massachusetts for state troops on April 5, 1775. In 1776, the Continental Congress appointed Thomas Jefferson and John Adams to draft new articles of war. A *Newsweek* story noted that these articles "made no provision for bail, indictment by grand jury, impartial judges or due process, and permitted cruel and unusual punishment, including branding on the forehead."[7] However, these articles did require the military to give up soldiers to the civil authorities for prosecution of capital crimes and crimes against the person or property of civilians. These articles of 1776 were accepted by the first Congress in 1789, though the provisions that conflicted with the Bill of Rights, with a few important exceptions, were dropped. Briefly revised in 1806, the articles remained in effect for nearly seventy years.

Jurisdiction of the military over civil crimes was not granted until the Civil War, when the first federal conscription law was passed on March 3, 1863. However, even this jurisdiction was not granted the military except in time of war, insurrection, or rebellion. In 1874, a complete codification of existing military law, consisting of 128 articles, was drawn up and remained in effect until 1916. Before 1874, the Articles of War had required that all men of the armed forces be given over to the civil authorities, both in peacetime and wartime, for trial in all cases except those concerning breaches of military discipline. In the articles of 1874, it was decreed that this requirement should apply except in time of war. From 1874 to 1916, the military was required to give over to the civil authorities all members of the armed services who committed civil crimes and all capital offenders.

The first complete revision of the United States military law in over 110 years came in 1916. This revision contained two crucial changes. First, the new articles stipulated that during peacetime murder and rape, if committed outside the geographical limits of the United States, could be tried by court-martial. Second, the new articles further provided that military jurisdiction be extended to such crimes as manslaughter, mayhem, arson, burglary, larceny, embezzlement, perjury, assault with intent to commit any felony, and assault with intent to do bodily harm. These articles of 1916 remained in effect until the Uniform Code of Military Justice was passed by Congress in 1950.

Certain procedural safeguards were written into this code which many claim were years ahead of such safeguards in civilian courts laid down by the United States Supreme Court over the next two decades. One of the major provisions of this Code, however, was that it made the military's jurisdiction over civil offenses virtually complete. The Military Justice Act of 1968 added more procedural safeguards and individual rights, but did nothing to restrict military jurisdiction over civil offenses.

Until very recently, the United States Supreme Court has tended to uphold military jurisdiction over civil offenses, and the few times the Court has concerned itself with military appeals, it has been quite specific. In 1866 the Court ruled that the power of Congress, in the government of the land and naval forces, is not affected by the fifth or any other amendment.[8] Likewise, in 1911, a majority of the justices ruled that "to those in the military or naval service of the United States, the military law is due process."[9]

However, on June 2, 1969, the Supreme Court ruled in *O'Callahan* v. *Parker* that a civilian-type crime must also be "service-connected" to be subject to court-martial jurisdiction.[10] The United States Court of Military Appeals had ruled in 1960 in *United States* v. *Jacoby* that it is "apparent that the protections in the Bill of Rights, except those which are expressly or by necessary implication inapplicable, are available to members of our armed forces."[11]

Since the O'Callahan decision, the United States Court of Military Appeals has held that service connection has been established in the following cases: crimes committed on post; crimes in which servicemen are the victims; crimes committed outside the United States; crimes involving use of military rank or status during commission thereof; crimes involving the use, possession, and sale of marijuana, narcotics, or prohibited substances of a similar nature; petty crimes; espionage offenses involving military documents; and crimes not cognizable by state or Federal civil courts.

IN FAVOR OF THE PRESENT MILITARY JUSTICE SYSTEM

> BAKER: *Men will go on as they are, subject to powers and authorities. And how are we to change* that *slavery? When it's of their nature?*
> CHIPMAN: *We try.*
> BAKER: *We redecorate the beast in all sorts of political coats, hoping that we change him, but is he to be changed?*
> CHIPMAN: *We try. We try.*[12]

Challenge the Uniform Code of Military Justice for not granting a soldier all the rights guaranteed by the United States Constitution, and the chances are the professional officer will reply, "Why should it?" For over a century now, military men have been fond of quoting Thomas Babington Macaulay:

> *The machinery by which courts of law ascertain the guilt or innocence of an accused citizen is too slow and intricate to be applied to an accused soldier. For, of all the maladies incident to the body politic, military insubordination is that which requires the most prompt and drastic remedies. . . . For, the general safety, therefore, a summary jurisdiction of terrible extent must, in camps, be entrusted to rude tribunals composed of men of the sword.*[13]

Dwight D. Eisenhower, in a similar vein, once remarked:

> *The Army was never set up to insure justice. . . . It is set up as your servant, a servant of the civilian population of this country to do a particular job, . . . and that function . . . demands . . . almost a violation of the very concepts upon which our government is established.*[14].

Even that celebrated watchdog of American constitutional liberties, former Senator Sam Ervin, remarked, "The primary purpose of the administration of justice in the military services is to enforce discipline plus getting rid of the people who think they are not capable of contributing to the defense of the country as they should."[15]

Despite the fact that many today would doubt the necessity for one kind of justice "military style" and another kind of justice "Constitution style," there are a number of individuals, both in and out of the armed services, who defend the Uniform Code of Military Justice as fair, just, and reasonable under any circumstances. As George Walton has put it, "Often I have thought that were I guilty of a crime, I would prefer a civilian court, and were I innocent, I would prefer a court-martial."[16]

The principal constitutional rights contained in the Bill of Rights omitted from the Uniform Code of Military Justice are the Fifth Amendment's requirement of a grand jury indictment in a capital, or otherwise infamous, crime; the Sixth Amendment's right to trial by an impartial jury of the state and district wherein the crime shall have been committed; and the Eighth Amendment's prohibition of excessive bail. In defense of these exceptions, it is said that the Fifth Amendment expressly excludes cases arising in the land or naval forces. Further, it is argued, not even Justices Hugo Black and William O. Douglas have contested a soldier's right to a jury trial for military offenses. As for the apparent violation of the Eighth Amendment concerning bail, the code provides that confinement of an accused pending trial and appeal shall be no stricter than is necessary to ensure his presence.

Proponents argue that the code offers the serviceman some remarkable safeguards: he must be advised before questioning of the offenses of which he is suspected; he has a right to remain silent and to be informed that anything he says may be used against him; he has the right to free legal counsel in general and special court-martials; in a punitive discharge, the accused may not waive his right to a qualified military lawyer; the accused may decide whether he wants to be tried by a military jury or by a military judge; if the soldier decides to be tried before a jury, he may have his case heard before a jury of officers, or one including one-third enlisted personnel; a soldier's trial must be speedy; the accused has a right to the personal appearance of all material witnesses at his trial, and all costs of bringing any witness from anywhere in the world for a court-martial are borne by the government; every serviceman convicted by a military court-martial is guaranteed automatic appellate review. In addition, any accused serviceman is given a copy of the full formal investigative report containing all testimony and all other material considered during the investigation: "As a matter of contrast in Federal grand jury proceedings, disclosure of any prosecution evidence is at the discretion of the trial court and, if allowed, may be subjected to being disclosed only on the first day of the trial."[17]

Reacting to the charge that too many military judges are under the influence of higher command, many experts reply that the code provides that any sentence which includes a dishonorable discharge or bad-conduct discharge or confinement for a year must be reviewed by a court of military review whose members report to no military authority except the Judge Advocate General. The decision of the court of military review may be appealed to the United States Court of Military Appeals, a civilian court appointed by the president. A general court-martial is presided over by a military judge who is not responsible to the commander who convened the court and preferred the charges, but only to the judge advocate general of the service concerned. Joseph W. Bishop concluded: "Taking one day with another, military judges are likely to be at least as honest and competent as the judges of state criminal courts."[18]

Finally, it is frequently pointed out that the code prohibits compulsory self-incrimination, double jeopardy, and cruel and unusual punishments. Although the average GI is not tried by a jury of his peers, Melvin Belli has remarked that military personnel make better jurors because, on the whole, they are more intelligent, more paternalistic, and cognizant of the accused's situation.[19] Those who would defend the code point with pride to the fact that the United States Supreme Court has never freed a military convict because of unfairness in his trial. James A. Mounts and Myron G. Sugarman stated that the Military Justice Act of 1968 places the judicial system of the armed forces ahead of most civilian jurisdiction in terms of judicial procedures and concepts of due process.[20]

AGAINST THE MILITARY JUSTICE SYSTEM

Many critics of the military justice system are fond of quoting Georges Clemenceau's statement: "Military justice is to justice as military music is to music."[21] Even partisans of the system admit that the principle — it is better that 99 guilty men go free than one innocent man be convicted — is hard to reconcile with Army discipline. A large number of individuals, both friends and foes of military justice, agree with Justice Black:

> *Traditionally, military justice has been a rough form of justice, emphasizing summary procedures, speedy convictions, and stern penalties so that military tribunals have not been and probably never can be constituted in such a way that they can have the same kind of qualifications that the Constitution has deemed essential to fair trials of civilians in Federal Courts.*[22]

Critics of military justice charge that military law is vague and that punishments meted out for the same crime vary so greatly as to be ridiculous. It is pointed out, for example, that Article 89 — "Disrespect Towards a Superior Officer" — is defined in the *Manual for Courts-Martial* as including "marked disdain, indifference, insolence, impertinence, undue familiarity, and other rudeness." Under this article, it would seem that the charge at issue would depend almost entirely on the capriciousness of the offended officer. There is little doubt that the vagueness of Article 134 of the same manual, which defines disloyalty as "praising the enemy, attacking the war aims of the United States, or denouncing our form of government," has created innumerable difficulties for the military itself since the beginning of the Vietnam war. Article 133 of the code forbids "conduct unbecoming an officer and a gentleman." Without further definition of this offense, the article makes every officer in the armed services subject to whim.[23]

Likewise, the punishments handed down by military courts are often arbitrary. Some servicemen have been sentenced to a simple discharge for refusing to wear a uniform, while others have served three years in prison for the same offense. Sentences ranging from several weeks to sixteen years in prison have been dealt out to soldiers for the identical crime — refusing to obey an order. Robert Sherrill pointed out: "Holding an antiwar bull session while in uniform on base has resulted in everything from an administrative discharge without punishment to ten years in prison and a dishonorable discharge."[24]

A second charge by opponents of the military justice system is that, despite the advantages theoretically open to defense attorneys for servicemen in courts-martial, defense attorneys in such trials are required to run an "obstacle course." In too many recent cases, it is charged, records have disappeared that were vital to the defense. The Army refused in the trial of the Presidio "mutineers" to take verbatim transcripts of the preliminary hearings and to supply the defense in later trials with transcripts of the earlier trial records. Records were "lost" by government agents in the Levy trial. At least one partisan of military justice agrees that the increasing amounts of "red tape" are making the task of the defense in a court-martial ever more difficult.

A third criticism leveled at the system is that the juries in many courts-martial are "stacked." F. Lee Bailey, the noted attorney has remarked:

> *In the case of military justice, the commander who orders the trial — a guy who is himself convinced that there are good grounds for conviction — selected the jury. And if the case is a heavy one, the officer in the jury sits there and reflects on his career in the military. He says "If I do justice, my conscience will feel better for a couple of days, but that son of a gun, the presiding officer, is going to remember me for years."*[25]

One defender of the military justice system, while claiming that few officers on a jury try to gain favor with their commanding officer, does admit that the right of a serviceman to have one-third of his jury composed of enlisted men is no great favor since "what he will probably get are first and master sergeants, who are likely to be rougher than commissioned officers."[26]

Enlarging on this criticism of the military jury system, many claim that the most pernicious factor at work in the entire system is "command influence."

> *The commander decides when and whom to prosecute. He controls the investigation of the charges and can overrule the officer who conducted the preliminary investigation. The commander can personally select the jury members from among officers who are beholden to him for favors, promotions and other career opportunities; he also picks the prosecuting officers and the military defense attorneys. Although the staff judge advocate is supposed to be a neutral administrator of portions of the trial procedure, he is in fact*

> *the commanding officer's attorney and, as such, represents the commander's wishes in all that he does.*[27]

The commanding officer is the first to review the trial record and the sentence. Émile Zola Berman, the famed trial attorney, has said: "The commanding officer never says 'I want to get this son of a bitch.' But he doesn't have to say it. The members of the court understand that they are here to convict."[28] One veteran lawyer with a great deal of experience in defending servicemen, Henry Rothblatt, said, "If the CO is out to get you, God help you."[29]

Edward F. Sherman summed up the power of the commanding officer in a court-martial:

> *It is a little like having a district attorney act as a grand jury, select the judge, both attorneys and the jury from his staff, and then review the sentence on appeal. The Code, in an attempt to preserve a fair trial, forbids commanders from influencing the action of a court-martial, but the possibility that the junior officer can banish the influence of his commander (who rates him and controls his assignments) is about as likely as a senator not being influenced by accepting large gifts.*[30]

Many charge that, if the soldier is found innocent by the jury, he still must face the hell that any hostile commanding officer can create for him. Such a serviceman may find himself in the position of Clevinger in Joseph Heller's novel *Catch-22* as the colonel asked him:

> *"Didn't you whisper to Yossarian that we couldn't punish you?"*
> *"Oh, no sir. I whispered to him that you couldn't find me guilty."*
> *I may be stupid," interrupted the colonel, "but the distinction escapes me. I guess I* am *pretty stupid, because the distinction escapes me."*[31]

Only in 1968 did Congress make two significant alterations in the Uniform Code. First, it created an independent field judiciary in each of the services from which experienced military judges are drawn to preside at general courts-martial. This order had the effect of taking away much of the influence of the commanding officer who ordered the court-martial originally. Second, Congress, encouraged by Senator Sam Ervin, gave each defendant the right to a trained military or civilian attorney in a special court-martial if physical conditions or military exigencies permit.

Nevertheless, many critics protest that there is a general atmosphere surrounding courts-martial of whatever variety which leads to the suspicion that the accused is guilty until proven innocent. They point to the fact that in 1969 the military held nearly 110,000 courts-martial and that military prosecutors ran up a 94 percent rate of conviction as compared to the 81 percent rate in federal courts for civilians.[32] One is again reminded of one of Joseph Heller's hapless characters in the dramatization of his novel *Catch-22:*

> CAPTAIN: *Why'd you steal it if you didn't want it?*
> CHAPLAIN: *I didn't steal it!*
> CAPTAIN: *Why are you so guilty, if you didn't steal it?*
> CHAPLAIN: *I'm not guilty!*
> CAPTAIN: *Why would we be questioning you if you weren't guilty?*[33]

Some critics further claim that command control still exists even after the 1968 congressional reform. Roger Priest, against whom two charges were dismissed by the military judge in his case, had the charges reinstated against him by the judge after the commandant of the Washington Naval District ordered such action.[34] Likewise, the cases of Green Beret Colonel Robert Rheault and Lieutenant William Calley are cited as further examples of command influence at a high level.

While advocates of the military justice system point out that certain cases may be appealed to the United States Court of Military Appeals — composed of three civilians appointed by the president to fifteen-year terms — at least one opponent of the system maintains that all the members of this court since its founding in 1951 have been staunch conservatives, and a number of them had backgrounds as officers in the military. Attorney Maryann Weissman goes so far as to suggest that the entire military justice system is merely a matter of one political philosophy facing another:

> *A military court-martial is an important political forum, but even more, it becomes a political contest in which brass measures its strength against enlisted men. Nowhere is the chain of command more direct than in a military trial of political significance. The Judge Advocate General's office,*

> *which appoints both prosecutor and military defense counsel, is a link in the same chain which authorizes the charges and the nature of the court-martial.*[35]

Perhaps the enormous importance of thoroughly reforming the military justice system has been stated best by General David M. Shoup (Retired), himself a former commandant of the Marine Corps:

> *Today most middle-aged men, most business, government, civic and professional leaders, have served some time in uniform. Whether they like it or not, their military training and experience have affected them, for the creeds and attitudes of the armed forces are powerful medicine, and can become habit-forming. . . . For many veterans the military's efforts to train and indoctrinate them may well be the most impressive and influential experience they have ever had — especially for the young and less educated.*[36]

In other words, the greatest danger which America faces from the present military justice system is that if military justice is corrupt, sooner or later it will further corrupt civilian justice.

MILITARY JUSTICE AND THE BLACK SERVICEMAN

> *Tonight, my friends — I find in being black a thing of beauty: a joy, a strength, a secret cup of gladness; a native land in neither time or place — a native land in every Negro face! Be loyal to yourselves: your skin, your hair, your lips, your southern speech, your laughing kindness — are Negro kingdoms, vast as any other! Accept in full sweetness of your blackness — not wishing to be red, or white, or yellow; or any other race, or face but this. Farewell, my deep and Africanic brothers! Be brave, keep freedom in the family, do what you can for the white folks, and write me in care of the post office. Now, may the Constitution of the United States go with you; the Declaration of Independence stand by you; the Bill of Rights protect you; and the State Commission Against Discrimination keep the eyes of the law upon you, henceforth now and forever. Amen.*[37]

Nearly seventy years ago, on the night of August 13, 1906, shots rang out in the town of Brownsville, Texas — shots which were to ring down the years as a reminder of the gravest miscarriage of justice in military annals. In the exchange of fire, a white bartender was killed and a white policeman was wounded, as the enraged whites of Brownsville sought to blame the incident on three companies of black infantry stationed at Fort Brown. Although their white commanding officer, Major Charles Penrose, found all his men accounted for in roll call as the shots were being fired, and though a rifle inspection the next morning revealed that none of the unit's weapons had been fired, all 167 men of B Company of the First Battalion of the 25th Infantry were ordered dishonorably discharged from the Army. The final orders to discharge the men came directly from Secretary of War William Howard Taft and President Theodore Roosevelt.[38]

U.S. Senator Joseph Benson Foraker of Ohio took up the case of the condemned black soldiers. Finally, in 1910, fourteen of them were found eligible to reenlist, but the others were denied such a right. John D. Weaver called the outcome "a triumph of military and legal cant over logic and justice."[39] As far as can be determined, only two men survive today who were members of B Company — Dorsey Willis and Edward Warfield. Willis, who has spent over sixty-six years trying to clear his name, said: "To take a person's rights from 'em is bad, you know. They had no right to eliminate me without tryin' me and findin' me guilty, but they did."[40]

In October, 1972, Secretary of the Army Robert Froehlke reversed Theodore Roosevelt's order and changed the discharges of the black servicemen from "without honor" to "honorable," pointing out that the Army had made a mistake in its treatment of the men. Warfield and Willis have since been compensated by the Army to the amount of $25,000, and a bill has been passed by Congress to grant funds to widows of the men involved.[41] Some small justice has finally been done, though the lives and reputations of over 150 men and their families have been damaged. It is doubtful whether the case would have had such a final outcome had it not been for the detailed book *The Brownsville Raid* by Weaver and the many months devoted to an investigation of the matter by Representative Augustus Hawkins of California.

A recent incident aboard the aircraft carrier U.S.S. *Kitty Hawk* threatens to become, in the minds of many observers, another

"Brownsville Affair." Between October 12 and 13, 1972, several disturbances took place between black and white sailors aboard the aircraft carrier, and the final result was that twenty-four men were held for trial — twenty-three blacks and one white. Different reasons for the disturbances were given by many witnesses at the trials. Whites maintained that blacks were "coddled" and given jobs for which they were not qualified. Blacks claimed that white crewmen had repeatedly assaulted blacks during the voyage before October 12 and that blacks were invariably given the "dirtiest" jobs to perform on the carrier.[42]

What cannot be denied in the *Kitty Hawk* incident is that the one white sailor accused was acquitted early in the proceedings and the accused black sailors received severe sentences. One case was felt to be a particular example of injustice, that of black sailor Cleveland Mallory, who was accused of riot and assault. Six witnesses failed to identify Mallory as the one who assaulted a white sailor, but a seventh witness, Michael Laurie, made a positive identification. On the basis of one man's testimony, Mallory was convicted and was given a bad-conduct discharge. It was later proven to the Navy board by a private investigator hired by the NAACP that Laurie had lied, and Mallory's conviction was overturned. However, Laurie was not charged with perjury, as would normally be the case, and the NAACP charges that this is just one more example of discrimination in the armed forces in favor of whites.

Another observer points out that the accused black sailors in the incident were kept in a lengthy pretrial confinement before they were brought to court. One black sailor, Arnold Petty, said:

> *My charges were not so serious as to warrant a general court-martial. I was tried by a special court-martial. The longest sentence I could have received was six months. Yet I was in prison for 65 days pending trial. The Navy, and especially Captain* [*Robert*] *McKenzie, had taken upon itself to punish me and the other men in the brig prior to a finding by the court-martial.*[43]

Even though the Congressional Armed Services Committee formed a special subcommittee to investigate the *Kitty Hawk* incident, that body took no testimony from any of the black sailors involved. The final subcommittee report, completely favorable to the Navy, was called a whitewash, for not only were there no black sailors interviewed but no black Congressmen served on the body. For this reason, Representative Floyd V. Hicks, chairman of the subcommittee, admitted that the report was "not credible in many areas."[44]

In 1974, another incident took place, this time on the aircraft carrier U.S.S. *Midway.* A large number of black sailors jumped ship in Yokosuka, Japan. These men charged that they and other blacks were discriminated against in duty assignments, promotions, and military justice. They alleged that Captain Richard J. Schulte handed out severe punishment at captain's mast. The ship maintained waiting lists for the brig because discipline was so severe, and beatings frequently occurred during incarceration, they further charged.[45]

One characteristic example of racial discrimination in the armed services is the case of Sergeant Elmore D. Smith, Jr., who only eight days before his discharge after four years of service, faced a court-martial. Smith, a veteran of Vietnam, had complained for several months about discrimination in the Marines. In the summer of 1971, he was transferred from Camp Lejeune to the Marine detachment at the Philadelphia Naval Base. Finally, on April 30, 1973, only a month before his discharge, he was ordered to get a haircut, even though his hair was well within the three inch restriction imposed on the Afro-style cut. He was also ordered to trim his beard, which he had grown, with medical permission, because of a skin ailment. Although Smith thought that he was being treated unfairly, he complied with the request. A misunderstanding arose concerning his reporting his compliance with the order to his sergeant-major. Colonel H. B. Wilson, Smith's commanding officer, when later asked if Smith's hair and beard conformed to Marine regulations, replied, "They are right now. That's not the point. They weren't before."[46]

Smith wrote a letter to Representative Shirley Chisholm of New York:

> *There is a dual justice system here. Many of the* [*black*] *men will not stand and be counted because they fear the consequences. I know because I am constantly being harassed because of my stand. There are constant threats against my stripe and my discharge and needless embarrassment in formation. All this because I dare to function as a black man.*

Smith maintained that Colonel Wilson told him that if he got a hair cut and removed his beard the incident would be forgotten. Even though he complied with the request, he was formally charged on May 7, 1973.[47] This incident, like so many others, tends to confirm the observation of Anthony Griggs: "Because it is the commanding officer — most often a white man — who determines the direction the legal action will take, the minority soldier is virtually at the mercy of a preponderantly white chain of command system."[48]

Of course, we could review the many cases in which black GIs — and other minorities — were at fault and deserved stiff sentences. Our concern here is with cases of questionable or gross maltreatment. Nor does the fact that white GIs are also victims of injustices make the results any less deplorable. A foremost task of the military is to recognize and prevent such treatment. The use of Article 15 and the dishonorable discharge have in many instances become unjust remedies for alleged wrongdoings.

ARTICLE 15 AND DISHONORABLE DISCHARGE OF BLACKS

> *But one thing, Ol' Cap'n, I am released of you — the entire Negro People is released of you! No more shouting hallelujah! every time you sneeze, or jumping jackass everytime you whistle "Dixie"! We gonna love you if you let us laugh as we leave you if you don't. We want our cut of the Constitution, and we want it now; and not with no teaspoon, white folks — throw it all at us with a shovel!*[49]

Of 919,349 discharges from the services in 1971, only 5,399 were bad-conduct or dishonorable. However, blacks received 1,092 or 25.3 percent of them — 19.3 percent of the bad-conduct discharges and 33.3 percent of the dishonorable discharges, despite the fact that blacks constituted no more than 15.1 percent of any branch of the military.[50] There seems to be something disquieting here when one remembers that by 1968, black deaths in the Vietnam war had reached 22 percent of all fatalities. Griggs states that "a picture emerges of a largely black and brown fighting force in the rice paddies and trenches, faced with the gore of exploding napalm, while the antiseptic, push-button corps of ranking officers and airmen is noticeably white-skinned."[51] Even more disturbing are figures from the Air Force: In 1972, blacks in the Air Force received 49.7 percent of the general courts-martial, 40 percent of the special courts-martial, and 37 percent of the summary courts-martial, even though they made up only 10.8 percent of this branch of the service.

What is often forgotten is the fact that the black GI suffers from two lesser forms of "other than desirable" discharges: "undesirable" discharges and "general" discharges. In order to give such discharges to soldiers, the military does not need to establish guilt through court-martial. In 1971, for instance, black servicemen received 11,871 out of 67,999 such discharges: 18.4 percent of the undesirable discharges and 16.9 percent of the general discharges. Those with less than honorable discharges are barred from veterans' benefits, including hospital and medical care, training and educational programs.

Congressman Carl Stokes of Ohio has proposed legislation to remove all classification from military discharges. He has pointed out that "a man may have to pay the rest of his life for a mistake he made at the age of 18 or 19."[52]

> *Their chief character defect was immaturity. Most of them have not committed serious crimes, but get into trouble often as a consequence of using alcohol or drugs, or both. Others suffered discriminatory treatment, went* AWOL *at a time of stress, or went berserk marking time the last few months in Vietnam. Also there are those who for mental or psychological reasons should not have been inducted in the first place.*[53]

Sherrill is emphatic: "If there is any one compelling reason for returning all Americans in uniform to the jurisdiction of the civilian courts, it is that the military courts make no allowance for the types and backgrounds of the people who come before them."[54]

The undesirable discharge is for all practical purposes as heavy a burden around the recipient's neck as is the dishonorable or bad-conduct discharge. It is a legal handicap because it can be given administratively, without any of the process due of the accused in a court-martial. Many experts charge that Article 15 of the Uniform Code of Military Justice is the culprit in many such discharges. No doubt it was largely because of the above figures and charges that former Secretary of Defense Melvin Laird commissioned an investigative group, called the Task Force on the Administration of Military Justice in the Armed Forces, to examine discriminatory practices in the armed forces. The task force began its study in April, 1972. It found that the military services are influenced by broad societal practices, including racial discrimination.[55]

Article 15 of the Uniform Code of Military Justice allows a commanding officer to by-pass the regular court system of the military: "He makes his own charges and backs the charges up with whatever he feels is the proper punishment."[56] The task force statistics point out that a

> *. . . disproportionate number of black soldiers have been issued out of the armed services by way of Article 15 after committing minor offenses. Conversely, white soldiers who have committed more severe infractions have been processed through Article 15 instead of being brought before a military court.*[57]

The task force further confirmed the suspicions of many by revealing that Article 15 "provides the greatest opportunity for the practice of racial discrimination. Many blacks feel — and our statistics confirm — that they receive prejudicial punishment disproportionate to their numbers in the military."[58] It should be pointed out here that over 1,000 enlisted men signed a petition to abolish Article 15. The petition was presented to Congressman Ron Dellums late in 1973. It charged that most Article 15 hearings involve minorities and men from the first four pay grades.[59]

As a result of the task force report, the Department of Defense revised Article 15 to include the following provisions:

1. The availability of adequate legal advice to an accused person prior to action by commanders authorized to impose punishment.
2. The opportunity for full presentation by an accused person of the case in the presence of his or her commander, to include, but not limited to, the right to call witnesses, present evidence, and to be accompanied by a person to speak on his or her behalf.
3. Each accused person to be advised of his or her right to appeal any nonjudicial punishment.

4. Imposition of punishment under the nonjudicial punishment Article 15 be stayed pending completion of any appeal filed.

5. Nonjudicial punishment proceedings to be opened to the public when requested by an accused except in those instances where military exigencies or security interests preclude public disclosure.[60]

In addition to the above changes in Article 15, the task force made separate recommendations concerning the military justice system in general. Among these were:

1. Eventual abolition of the summary court-martial.

2. Increase of the authority of the military judge to include suspension and deferment of sentences.

3. Random selection of court members rather than through the selection of the officer who refers the case to trial.

4. Addition of further peremptory challenges, with the defense having a greater number than the prosecution.

5. Enactment of a specific article banning discrimination.[61]

The Congressional Black Caucus, but for whose constant concern the task force might never have come about, made the following important additional recommendations:

1. That the Uniform Code of Military Justice be amended so that discrimination be an offense considered punishable by court-martial just as any other form of misconduct.

2. That legislation be introduced to eliminate all punitive discharges and to establish in their place a certificate of service. Each person upon completion of 181 days of service would be eligible for any veteran's benefits provided for under current law. The military would continue to have authority to discharge a soldier and to punish him for violations of their law. However, the military would not be able to mark an individual for life. If the service decided to discharge an individual after this given period of time, it would not be able to strip him of his benefits.[62]

Perhaps one of the most meaningful recommendations of the task force was that a more effective orientation program concerning the military environment, its laws, practices, and differences from most civilian environments be provided early in the enlistment of every service member. The report further recommended that periodic uniform refresher courses of instruction in military justice be given to all servicemen. It would appear that the Navy, at least on some bases, is attempting to comply with the latter recommendations. In 1973, the San Diego Naval Base ordered that boot camp training be raised from seven to nine weeks so that recruits may be better instructed on Navy grievance procedures. The Navy has stated that it plans to use the additional two weeks for more instruction in military law, Navy customs, and shipboard procedures.[63]

The Congressional Black Caucus report sums up the problems which the military justice system faces as well as any document of the past few decades:

> *As one reads the Uniform Code of Military Justice, one realizes that it is an arm of discipline of the commander. It is not set up to resolve the disputes between those in authority and those who are supposed to be commanding. It is a tool by which the commander compels the men to do whatever the commander wishes, rightly or wrongly, and that kind of system is not a law system. It is certainly not justice.*[64]

Colonel George Walton, a veteran of the United States Army and one of its staunchest defenders, stressed that one of the most effective moves the Army has made in recent years has been the involvement of more and more nonwhite and white enlisted men and officers in racial seminars on their respective post. The purpose of these gatherings is to "seek out, identify and eliminate the causes of racial tension or unrest in the Army," and to "ascertain ways and means of developing positive measures whereby racial tension within the Army can be eliminated and equal opportunity and treatment of all military personnel can be assured."[65]

No responsible discussion of the military-justice system can leave any individual in doubt about the seriousness of the lack of communication between officers and personnel in the lower ranks. If our military service and its justice system are to survive, swift and concrete steps must be taken to insure that channels of communication are opened up between the system and the men and women responsible to it.

Justice is More Than a Word

Brian L. Ford,
formerly Captain
Social Action Officer, USAFE

CASE HISTORIES

The following individual and admittedly incomplete case histories are true. Except for changes in location and individuals' names, the facts presented are, to my best assessment, a fair representation of actual events. Relevant concerns are presented and discussed. The Air Force was chosen simply because it embodies a society directly experienced by this writer and because the programs currently in existence allowed fuller investigation of the facts surrounding each case. Furthermore, as Lieutenant Colonel Renfro of Hq. USAF Equal Opportunity and Social Actions remarked, "The more open the society is to change, the more volatile and open the change will become." The development of social actions offices which exist at every major base, — including the fields of race relations, drug abuse, equal opportunity, military personnel treatment, and alcoholism, — has led to the emergence of issues and concerns long hidden. The expectation of some action being taken has led to a greater confidence in the possibility of solution and hence to a greater honesty and forthrightness in presenting problems and concerns by its abused members.

Specifics of military service, operation administration, and so forth, are, for the most part, explained. It might be of help to the reader to review the rank structure of enlisted personnel (various private, corporal, and sergeant designations) and officer personnel (from lieutenant to general), as well as the social system of control (command being roughly comparable to a combination of administrative, judicial, and operative authority within the constraints of the regulations that govern a specific operational span of authority). Hence, murder might be tried or "court-martialed" by the Air Force, depending on the specific base in question, while disciplinary authority for job-related violations remains in the hands of the specific commander. The commander might counsel (and warn) an airman concerning his errant behavior. If that misbehavior continues, he has the option to make further reprimand a matter of record, as a last resort. If an individual airman refuses to accept a commander's conclusion, he can elect to be tried by court-martial, with the possibility of conviction, subsequently recorded as a violation of federal statute. Consequently, an individual who feels strongly that his punishment is either out of proportion to his offense or that he is being singled out — but who committed the offense nonetheless — is placed in the following dilemma: should he question the action and risk federal conviction or accept the single commander's possibly biased view? The only check on this system is that the immediate superior commander has the authority to review the case, which therefore rests on his sensibilities. Though punishment is often deemed too harsh and is reduced, the efficacy of imposing punishment at all in the case of an admittedly guilty party, is seldom if ever questioned.

This type of review is now being undertaken by the social-actions and legal offices in an effort to force commanders to function, at least in part, on precedent, using statistics and impartial investigation to reveal bias where it exists. In many cases, infractions occur that are not punished. The decision to punish should be made against this backdrop. The power to act remains in the hands of the commander; the power to embarrass being the retaliatory weapon of outside investigating offices.

The cases which follow have been presented in such a way as to avoid dehumanizing what was, in each case, an agonizing human situation. However, facts are not misrepresented, nor are characterizations distorted. Reality must be approached both subjectively and objectively.

The military is, in many ways, a microcosm of the larger society. The functions of authority, power, status quo, security, and shared responsibility are accented, along with a heightened mobility. The military is both more and less responsive to human relations, and perhaps an ideal forum for examining the elusive intra- and interpersonal concerns of the human being.

THE WASHROOM AFFAIR

Very early one morning, shortly past midnight, the Security Police were summoned to the Aircraft Maintenance Squadron barracks because of a disturbance, the noise disturbing the adjoining barracks personnel some two hundred feet away. The police arrived and went to a second-floor washroom filled with a hodgepodge of more than forty angry white and black airmen. The washroom, having been designed for the personal hygiene of up to eight men, was understandably crowded. In fact, several carloads of police reinforcements were required to unpack the area.

Eight black and twelve white airmen (one of whom resented the term "white," insisting that he be listed as "Chicano," and the other protesting his inclusion as "Chicano," preferring the original "white" designation) were incarcerated overnight in the brig. The jailhouse had two rooms in which the prisoners were segregated by the above classification despite prohibitory military directives to avoid using racial classification for incarceration actions, simply because on that particular night when any blacks were housed with whites or vice versa, the free-for-all began anew. The responsible squadron commander arrived on the scene sleepy-eyed, and after a confused recounting by the Security Police of the night's events (muddled somewhat by the diametrically opposed versions emanating from the two cells, and the suspicious presence of the odor of alcohol coming from the same room), he elected to leave the participants in jail overnight.

The next day, the entire base legal staff was sequestered in the brig taking chain testimony from the opposing sides. Despite the conflicting accusations, a fairly clear picture of the preceding week's events and of the melee in the second-floor washroom appeared.

Some weeks before, the commander in a visit to the barracks discovered a number of "anti-American" posters displayed on the walls of rooms belonging to various individuals, and coincidentally black airmen. Among the slogans of these posters were "Power to the People," "Freedom by Whatever Means Necessary," "Don't You Love Pigs?" (depicting several porcine members of some unfortunately immortalized police department), and "Black is Beautiful" (with a clenched fist). The commander had ordered these posters removed under the threat of confiscation. He included, in this order, removal of a number of nudes adorning the same walls. Oddly no posters of nudes were removed from the rooms of white airmen, several of whose rooms were decorated with large Confederate or state of Georgia flags (in which the Stars and Bars play an important part). A complaint was lodged with the commander by several of the affected and offended black airmen, who were informed that the display of legitimate flags was within regulations (the Confederate flags were ordered down in an effort, the commander later said, at reconciliation.)

The next day, a large banner in red, green, and black — colors of Afro-America — adorned the wall of one of the previously inspected rooms of two black airmen. Word reached the ears of the first sergeant speedily, and that afternoon the banner was included on the proscribed list. The airmen refused to remove it.

They were called in for counseling. They belligerently refused to remove the banner. They were given a direct order to remove it. They replied that it would remain until the Georgia flag came down. The black airmen were each given a letter of reprimand, and the order was repeated. They grudgingly complied but called a soul meeting for that night.

While the meeting took place, at first quietly and then with greater gusto, the melodious strains of "Dixie" were heard in the hall outside, emanating from the room of one of the white participants. The door opened to an irregular march of squadron whites, behind the Confederate flag. They then began to march up and down the hall. Miraculously, the blacks departed that evening without confrontation.

The next morning, new complaints were registered with the commander, punctuated by comments by the black airmen indicating that "if [he] didn't do something, they would." The commander visited the barracks that night, and was greeted by calm and silence. He utilized the occasion to enforce the earlier ban on pin-ups in one of the blacks' rooms, the walls of which were now resembling pincushions, and then he retired for the night.

The following day, there was a mild dispute over priorities at the morning sinks. One particularly vocal white, who used the term "nigger," later made a visit to sick call for minor cuts and bruises. That night after work, the Confederate parade reformed, blocking the entrance to the complainant black's room, inviting him and his "chickenshit nigger buddy" to come out with fists raised. While this lively repartee ensued, the night shift finished work, and the remaining blacks, accompanied by the Chicano, (who was some hours later to insist on that designation), topped the stairs, and battle was joined, despite the peacemaking efforts by the Chicano.

Several chairs were removed from the rooms, along with the Mexican-American (who preferred the term "white") and wastecans, ashtrays, and other implements were used as weapons. Later, when the Security Police had managed to calm the participants, two whites were hurried to the emergency room, one for removal of pieces of glass ashtray embedded over one eye and the other for an indentation in his head, roughly in the shape of a chair leg.

Final disciplinary action resulting from the incident provided three of the black airmen and the Chicano with early adminisdischarges from the Air Force (a nomenclature destined to remain with them permanently), and the remaining five blacks received disciplinary punishment, Article 15s, and fines or reduction in rank.

No black airman testified in his defense, except to recount in a sworn statement that he had been called "nigger." No black airman admitted to using a weapon of any kind or testified against any other black airman. None of the three discharged airmen contested his separation, despite legal explanation of the seriousness of the type of discharge that could be expected if he did not do so. Not one of the white airmen was disciplined.

Several days after the disciplinary action, the commander had a rap session with all his men, announced on the squadron bulletin board and at morning formation in advance, "to air grievances." Two blacks attended. As a result of that meeting, new curtains and chairs were installed in the barracks lounge and several individual shift changes were made. Later that night, the commander's office was burned.

A thorough investigation failed to reveal the arsonists, and no further acts of violence were reported. The three black airmen who received action were separated from the service within seventy-two hours of the washroom incident. Within two months, five of the remaining blacks had been transferred from the base or to other units.

Some time later, the banned Confederate flag returned, hung in its original position. No one complained.

ANALYSIS

An irony exists in the punishment of so many blacks and so few whites for what was plainly a joint altercation, precisely at the time when the Air Force was attempting to achieve equal treatment and gain credibility in the effort. It was a manifestation of the lack of credibility on the part of the black participants, who expended so little effort in proving their case. When interviewed by a black member of the Social Actions Office, each participant readily divulged his version of the washroom saga, complete with quotations, personal identification, and moral justification for his involvement. Nonetheless, each individual later refused to testify in his own defense. The rationale was clearly expressed by the occupant of the "poster" room: "Brothers already understand what went down, and the Man ain't going to believe us no way; so I ain't admittin' nothin'."

Given the historical image of the white (and often corrupt) policeman in the ghetto and the jurisprudence evident only from a defendant perspective, there can be little rational expectation of collaboration by young blacks when authority is perceived to be the same (white, middle class, conservative). This perception was reinforced as each successive washroom battler was found guilty without the benefit of prodefendant data which was seen to be self-evident by the blacks. Furthermore, a cultural difference was evident in the selection of weapons, since "coming out to fight" in the ghetto carries greater life-and-death significance than in suburbia. The blacks viewed the hospitalization of several of the whites as proof of their just determination. Needless to say, the judiciary viewed it as assault and battery. Rather than proof of the rationality of the system and its logical objectivity in coolly considering all facts presented, the decisions of guilt served to add to the burning heap of racial misunderstandings.

Another significant issue arose in the consideration of the posters, the focal point of the initial misunderstanding. Though some placards ordered removed had undeniable political implications, those reinforcing the status quo ("America, Love It or Leave It," an old Confederate soldier saying, "Forget, Hell") were not subject to the same judgment. This undercurrent of implicit morality was not unnoticed by the black community, who protested loudly the unilaterality of the initial actions and subsequent punishments, not their judicial rightness.

The grouping around a racist symbol seemed, on investigation, almost a pretext, since most of the white airmen involved were neither from the South nor had they previous records of racist involvement. This observation was reinforced by the young blacks' strenuous efforts to attain greater recognition on base rather than through racial complaints within the squadron or the barracks.

In *The Adjusted American,* the Putneys hypothesize that hatred is redirection of residual internal questioning still unresolved. Here the spontaneous effort to incite confrontation appears almost cathartic, an effort to demonstrate that vocal and activist young blacks warrant suppressive reaction. Their symbols might be seen as virtually archetypal; the misty spiritual and human concerns clearly have not been resolved by the society and are evident in the specific enunciation of troubling ethical questions as they manifest themselves in the black experience.

The white airmen, as "true believers," described by Eric Hoffer, found their spiritual solace and resolution in group submersion. It is my contention that in social systems rewarding conservation of the status quo, there is a large potential market for reactionary acting-out. Thus the active Confederate flag carrier drew an avid following from the disturbed "in system" membership of the aroused squadron, a membership troubled by the threat of change. The colloquial paradigm is, "I came in the service for a little stability, and now the blacks start this!" Furthermore, the deviancy seen in the vocal blacks was a projection of very real, continuing anxiety within each participant.

The blacks, by their activism, made this white-dominated haven less comforting, and they reaped the resulting whirlwind.

ABDUL JONES

Abdul was a staff sergeant and the head of a nuclear-weapons loading team. The restrictions on entry into the nuclear maintenance field were such that an entire regulation existed covering personal behaviors considered to be sufficiently "abnormal" to warrant disqualification. Abdul had been given a medal in Vietnam for his bravery and another for his "professionalism." He intended to remain in the Air Force as a career, saying, "It's not perfect, but it sure beats the 'world' [referring to civilian life at home in the United States]."

He was black, twenty-five, and about to be court-martialed.

Several months before, Abdul had received notice of one of the ubiquitous and trivial duties falling to junior NCOs, manning the squadron phone all night on the very day the duty was to be performed. Abdul asked the first sergeant to be relieved, since he was attending university classes at night and would miss his midterm examination. He was informed that he could be relieved only if he found a replacement himself. Abdul pointed out that his duties were to begin within one hour and that most of the squadron had already been relieved of duty and had sped for home. The first sergeant was unmoved.

Abdul luckily discovered a replacement and returned to the squadron administrative office to ask to see the commander. The first sergeant demanded to know why. (This squadron had a published open-door policy, allowing a man to see the commander when he wished without prior screening or permission.) Abdul refused to explain. The first sergeant said that Abdul would have to wait because the sergeant "should have a chance to solve the problem himself." Abdul replied, "You done blown your chance, baby," at which point he received the not undisquieting words that his comments would "cost him a stripe."

Several moments later, the commander appeared. Abdul insisted on speaking with him. The first sergeant said, "This man has been causing trouble all day, sir. He won't even let me help him and he has been insubordinate." The commander elected to see Abdul the next day.

With the race-relations officer in attendance the following morning, the meeting with all affected parties revealed that the first sergeant had made a mistake — that the duty roster had not been published in time and that no disciplinary action would be taken against Abdul. The matter was closed.

The following day, Abdul's section received a spot inspection, the first in more than three months, and a number of "discrepancies of a serious nature" were uncovered. The section was put on additional duty, and Abdul received a letter of counseling for his "attitude problems" that "were evident" in his protests over the inspection.

Abdul took the issue to the Human Relations Council, a body composed of representatives from every unit on base. He asked that the council review the files and records of spot inspections in the squadron to determine if the action taken was justified. Abdul was called into the commander's office and asked if he "liked it here" and "what he expected to achieve in stirring up the unit?"

Later that month, Abdul attended the race-awareness seminar, a semiencounter approach acquainting Air Force members with the real discriminatory problems still extant. He was eloquent in his recitation of his personal life experiences and in his questioning of a female captain, who stated that racial problems were "none of my affair." The session was attended by Abdul's supervising officer, who defended the captain.

Subsequently, that week Abdul was not allowed to participate in planning of a base-wide minority cultural activity, being restrained by his supervising officer owing to "work requirements." He asked to see the commander, who spoke to him for less than a minute in the hallway outside his office, finishing the interview with the remark, "You had better start doing some work around here, Sergeant, if you expect special favors." (All base personnel had been authorized by the wing commander to participate in the planning unless "critical job requirements" made it impossible.)

The following month, Abdul's car was stolen and left on the far, vacant side of the base, stripped. The oil had leaked out, and the engine was frozen. Abdul received a parking ticket, which arrived on the desk of the commander. The resultant counseling session yielded another letter of reprimand for insubordinate language and failure to obey a direct order (to acknowledge receipt of the ticket).

Abdul's efficiency report came due that month, and he was rated slightly below average. In the inflated world of military efficiency reports, anything below exceptional is a "failure," and there was little question that the rating would harm his further chances for promotion, which, up to that time, had been exceptional (given his accelerated rise through the ranks to staff sergeant). Abdul appealed the rating.

While the appeal was pending, he was ordered to a base in Turkey on temporary duty. Though this duty was a regular part of his assignment, Abdul was not scheduled to depart for another month. However, an illness in the unit required the presence of an "experienced and outstanding" crew chief since the urgent mission could not be left untended. In the throes of appealing his substandard rating, Abdul departed to Turkey to fill this urgent and highly technical position, the "outstanding NCO" assigned by the same commander who had endorsed his efficiency report.

One week before his return, an unusual occurrence required the personal attention of the wing commander to Abdul's section in Turkey. The senior master sergeant in charge of overall munitions maintenance at the base, having returned from the NCO club in the section's pickup truck, was confronted by Abdul, who asked to know why he and two of his men had been left on the far end of the runway some three hours before. They were known to be without transportation, and no one had returned to share the vehicle with them so that they could proceed to the mess hall for lunch. Abdul was told that the vehicle was "busy."

The senior master sergeant then noticed that the surrounding ramp area needed sweeping (a task not reserved for munitions specialists) and suggested that Abdul take appropriate action with a broom. Abdul saw this as deliberate harassment and evasion of the question, and he made his feelings known to the sergeant. They were joined by Abdul's supervisor, a technical sergeant, who received from the senior a request that the ramp be swept by Abdul and his two men. He was told that they had "refused an order." The tech ordered Abdul to proceed with the sweeping. Abdul attempted to respond, and the tech sergeant said, "I don't want any of your complaining, boy." He tossed a large broom at Abdul.

Abdul, restrained by one of his men, left the hangar. He and his men traveled a mile on foot to the door of the major who supervised maintenance on the line. Abdul made his complaint known and was assured of an immediate investigation.

The next day Abdul returned at the hour designated by the major and received notice of pending Article 15 disciplinary punishment because of complaints from the senior master sergeant and the tech sergeant. Abdul's complaint was missing. He asked about his omission. The major replied that his "investigation" revealed no evidence to support Abdul. He was dismissed.

Four days later, Abdul returned to his home base with three statements from witnesses attesting to the accuracy of the story recounted above. He drafted a complaint which was delivered to the wing commander personally. The base Social Actions Office was asked to investigate the incident. It did and supported Abdul. Abdul refused the Article 15 and insisted on a court-martial.

A day passed. The charges were dropped. The wing commander personally (though verbally) reprimanded the commander of the base in Turkey. The major was counseled. Abdul was summoned before his commander and informed that all action had been dropped. He asked about action on his original complaint. The commander replied, "*All* action has been dropped. The sergeants have been counseled." (This prevented further action from being taken.)

Abdul applied for release from the Air Force. In the final weeks of his duty, he was accused of failing to button his coat, having been identified by the license number of his abandoned car, next to which he (allegedly) was standing. He denied the charges but was placed on the administrative control roster of "personnel warranting close observation" and was warned that any further violations he might commit would be punished immediately. He was warned that he was being watched. He remained on the roster until the day he departed military service. He did not make the Air Force his career.

ANALYSIS

It is evident that Abdul's efforts were expended in a quest for a reality determined on a philosophical and impartial plane rather than from a realistic viewpoint. His final words to his counselors on base were, "If this can happen to me, it can happen to anyone. I aim to stop this thing now." Though cleared of the "unzipped coat" charges, he left the service in much the same condition as he entered it. No further action was taken against the senior NCOs or the major who pressed unsubstantiated charges against him. Despite his efforts the matter, as predicted, remained closed.

Abdul's concern for "rightness" transcended his initial career defensiveness, and, had he remained in the Air Force, he would have progressed in grade much more slowly since he had received marginal effectiveness reports for his final volatile months in service.

The major himself sought escape from the mobility of action and freedom of decision afforded him by military regulations by reflexively supporting the senior NCOs. To support Abdul, even by an investigation, was to call into question his own failure to recognize an inherent problem prior to its manifestation in specific incident. As Eric Fromm commented, the very freedom to act across a broad spectrum of possibilities, having potentially deep social implications, can be immobilizing to one unprepared for such action. The major opted for the safest alternative. Abdul had, in his choice of words and by the definition of the ranks of the participants, violated one of the canons of military life. He had been insubordinate. He thereby became readily available for punishment, which later could be referred to as "action taken" in the case.

Abdul became progressively more sensitive to his treatment, more alert to the subtle ramifications of duty assignments, promotions, counseling, and so on. This heightened awareness, coupled with his new status in the minds of unit authorities, gradually altered his conclusions concerning his future. The more sensitive he became, the more irritated he was by the insensitivity of others. The more aware of his own identity he became, the less tolerant he was of abuse of that sense of self.

When he decided that he had reached a certain heightened sensitivity and had achieved the recognition of sufficient numbers of incidents aimed at altering his heretofore "outstanding" behavior, the magnitude of his disenchantment surpassed his dreams of career, his plans for his family, and his compatibility with the Air Force. He separated. He "couldn't take no more" either.

SMITTY, THE CAPTAIN, THE COLONEL, AND THE COURT-MARTIAL

The base commander was lecturing the members of the Human Relations Council, in the earlier days of its formation, on the necessity of *their* being fair in investigating complaints of discrimination and also concerning their mistakes and their resultant image of radicalism and irresponsibility. The colonel was warming up to his subject, confident that he was calming the wilder elements and making his base safe for normalcy and the backwaters of status quo, far away from the embarrassing eddies of change. Recently the council had uncovered significant evidence against the assignment policies of a senior officer on base and, when no action had been taken, had written a number of influential congressmen, causing some embarrassment to base officials, who had to defend their inaction in the case.

The colonel was well into his chapter on "unsubstantiated claims and finger pointing" when he was confronted by an unusual figure — a crisply dressed airman. Last June's Airman of the Month, in fact. Hair trimmed in a "natural," unquestionably short of Afro status, arms free of identifying wristbands, shoes resonating slightly, stood Staff Sergeant Jeremiah Smith, better known as Smitty Number Two (the number-one designation on that base reserved for another Smith present that day and successfully playing Clarence Darrow to the colonel's introduction of exemplary evidence, much to his discomfort).

Smitty Number Two. Now here, thought the colonel, was a reasonable voice, at last. "I'll give you a substantiated claim," said, Smitty, "mine."

Three months before, Smitty told the council, he had been called in for "counseling" by his senior NCO supervisor while on temporary duty with his unit at a remote base. Present when Smitty entered were his immediate supervisor (a technical sergeant, who had told Smitty on arrival, "I don't expect you *boys* to see any trouble while you're down here, see?"), another tech sergeant from headquarters (his friend), and the senior master sergeant, overseer of the entire work area. They reminded him that this was strictly an informal session and that he could feel free to speak as he wished. The meeting was held to "clear the air."

Smitty was then informed that his "attitude" left something to be desired, that he didn't seem to be a part of the "team," that the counseling would "go on the record," and that unless he agreed to "shape up," further action would have to be taken. Smitty asked whether an example of his failure to participate in this "team" effort could be given him, since as a recipient of the previous Airman of the Month award, he was under the impression that his team spirit had satisfactorily evidenced itself. The technical sergeant began informing Smitty that if he "talked back" he could expect harsher action, but the senior master sergeant cut in, giving "this morning's episode" as an example. Smitty had refused to sweep off the aircraft parking ramp area. Why?

Smitty replied that as a staff sergeant he was a supervisor and that the job was not part of his section's, or even his unit's, responsibilities.

While the two sergeants conferred on the relevancy of this comment, the captain entered the office, passing between the conferring NCOs, and sat down at his desk. Smitty had picked up a magazine and was looking at the cover while the point of contention was being discussed.

Suddenly the captain spoke. "Put that magazine down. You pay attention when someone talks to you. You hear?"

"I'm sorry, sir. I didn't know you were talking to me," Smitty replied.

The captain was on his feet. "That will be enough of your lip. One more remark from you, and you can expect disciplinary punishment." Smitty put down the magazine, and shortly the meeting ended. The chief NCO decided that a mistake had been made. Smitty didn't have to sweep the ramp.

The next day Smitty received notification that he was to receive administrative punishment. He was on the plane back to his home base. Immediately on arrival, he requested an appointment to see his commander to confirm the notification.

It was confirmed, notwithstanding the fact that the captain had given only a warning. Smitty was to be punished for insubordination to an officer and to an NCO. He now could elect to accept whatever punishment the commander chose to impose, within certain broad limits of monetary fines and reduction in rank, or he could insist on a court-martial. The only drawback to choosing a trial was the possibility of receiving a conviction, which would become a matter of permanent federal record. The "company punishment," though it might impose harsh penalties, carried no implication of permanent record. The court-martial guilty verdict was a federal conviction.

Nonetheless, so overwhelming was the evidence in his favor that his lawyer advised Smitty to go to trial, despite the risk. At that trial, the facts appeared as they have been recounted. The senior NCO even testified in Smitty's favor that he had seen no insubordination. The technical sergeant admitted that Smitty had been told that the informal meeting was "off the record." The captain admitted that Smitty had obeyed his order to replace the magazine, though he insisted that Smitty's "attitude" in picking it up was insubordinate. Smitty was convicted.

He received the maximum penalty: forfeiture of half pay for two months, two grades' reduction, and thirty days' confinement.

Smitty appealed to the base commander (now standing silently in front of the council), who supported the punishment.

The matter was referred to the wing commander for review, who was so incensed with the proceedings that he suspended all the punishment and verbally reprimanded the captain. He did not overturn the conviction, however, somewhat precedent-setting, because Smitty "did pick up the magazine, after all." He admitted to Smitty that the case should never have been pursued, even at the lowest level. However, the conviction remained.

On leaving the courtroom, the staff judge advocate, a lieutenant colonel and chief impartial legal adviser to the commanders, was overheard through an open door by the entire courtroom to say, "Anyone who insists on a court-martial deserves to be found guilty for costing the government so much time and money. This ought to teach him a lesson." The lawyer was *verbally* reprimanded as a result of the ensuing furor. No further action was taken.

Smitty finished before a hushed council. Silently the colonel sat down. Several of the members were carefully examining the carpet. The fierce Smitty Number One slowly broke his pencil.

Then, at the door of the council room he added, "I understand, though. Everyone makes mistakes. At least you people listened to me."

Smitty, a career enlisted man, withdrew his reenlistment papers. He scheduled himself to return to civilian life, with his conviction, in two months.

ANALYSIS

Operant from the first moment of meeting between the senior master sergeant and Smitty Number Two were hidden assumptions concerning their interpersonal relations.

Smitty was invited to be open. He was. Hidden was an assumption that openness had limits and that candor was to be constrained. As a black, Smitty had interacted in a community that placed a premium on "telling it like it is" among contemporaries. He who had his facts "together" or who was most adept at presenting them, won verbal encounters by acclamation. The sergeant's invitation was interpreted by Smitty to be an opportunity to take part in a candid interpretation of duty requirements.

The sergeant, on the other hand, assumed that "counseling" would take place, during which Smitty would reveal whether or not he had refused to comply with the order and, if so, then would receive advice and direction from the sergeant to avoid recurrence, "off the record."

The captain assumed that direction was being administered by a higher-ranking enlisted man to a lower-ranking one. He assumed insubordination in Smitty. This is part of the "Blacks are pathological" syndrome used by many whites to justify their brash and reflexive reactions. Blacks are seen as intransigent, incapable of rational behavior, and reflexively resistant to reasonable interchange in "normal," that is, white, society. Consequently, the captain behaved defensively, visualizing a potential threat to *his* position.

Furthermore, proceeding on the assumption that the society functions with reason, elevating those capable of performing more complex and demanding tasks, the captain assumed that the senior master sergeant was right in the (perceived) interchange, and the senior master sergeant assumed that the technical sergeant was correct within that society's terms. A dilemma arose when it became evident that, even in those terms, Smitty was correct. The conversation ceased to allow time for the senior master sergeant to consider this phenomenon, and Smitty nonchalantly picked up a magazine, sensing impending vindication.

His comment to the captain, therefore, was both an honest response and a statement of defense, the spectre of another — and doubtless biased — opinion entering the scene and diverting a well-earned victory. This, of course, prompted his reaction.

The colonel, in the midst of a human relations council meeting — where certain values of equal opportunity, personnel treatment and honesty were supposed to be shared, — was revealed to hold far different values. A crisis of confidence in its most crystalline form, the revelation of this duplicity justifiably ended his participation on the council. Later his reactions against certain of the council members involved in other incidents on the base resulted in his removal by the wing commander. (Once again, though, this reflects the awareness of an individual commander, not the responsiveness of the society.) A similar occurrence under the aegis of the same base commander before him had also resulted in disciplinary action.

One feels shamed by Smitty's closing remarks to the council: "Everyone makes mistakes." His sensitivity was born of human exposure to the vagaries of life and an appreciation of the irrationality of hatred, prejudice, and fear in a society. From that understanding stemmed a strong sense of emotional identity not dependent on the arbitrary application of justice and reinforcement of that society. He accepted the reality of the military society's actions without accepting their validity. He accepted the humanness of that society's members, apart from their participation in it. His available alternative behaviors were rebellion or submission, and yet he rebelled only by example.

To feel is to be human, and to be human is to feel, among other attributes. Smitty's coarse and affective background had prepared him well with a full repertoire of feelings, and his strength of identity enabled him to experience them without submitting to them involuntarily. He behaved conscientiously and sensitively because of those attributes and without the pressure of enforcing society.

In other words, Smitty was an independent, caring human being. Perhaps my emotional reaction to his final words stemmed from empathy of recognition.

QUESTIONS FOR FURTHER CONSIDERATION

1. Should the entire structure of the punishment system of the armed forces be reviewed in light of the new all-volunteer Army?

2. Should workshops be held preliminary to induction to insure that recruits are aware of all the benefits and liabilities in the military justice system?

3. Will the large number of women volunteering for the new army necessitate any realignment of the military justice system?

4. Should minority groups be encouraged to prepare themselves for a legal career to be pursued in the armed forces owing to the increase of their numbers in the all-volunteer armed forces?

5. Should those in decision-making capacities, as they pertain to legal justice, reflect the racial composition of the armed forces?

6. Should the number of enlisted personnel serving on juries trying noncommissioned defendants be increased?

7. What significant changes have occurred in the military justice system since 1974.?

part 5
The Use and Abuse of Alcohol and Narcotics

The first step makes me free, the second makes us slaves.

— *Goethe*[1]

Alcoholism and drug addiction take an ever-increasing toll each year in the United States, in terms of both human lives and money. A recent report of the United States Department of Health, Education, and Welfare to Congress stated that at least ten million people in this country are alcoholics or "problem drinkers."[2]

The annual cost of alcohol consumption to the American economy, says the report, can be broken down into various categories: $9.35 billion in lost production of goods and services; $8.29 billion for health and medical care; $6.44 billion in motor vehicle accidents; $640 million in alcohol programs and research costs; $2.2 billion in welfare payments; $500 million in criminal justice costs; $135 million in social services costs; and a considerable portion of the $4.5 billion reported last year as fire losses can be directly or indirectly attributed to alcohol use. Perhaps the most tragic statistic of all is that at least one-third of all homicides and one-half of all traffic deaths are linked each year to drinking.[3]

Figures on the use and abuse of drugs other than alcohol, while not as reliable as those on alcohol, are nevertheless frightening: as many as 24 million Americans have used marijuana; 19 million Americans are users of such stimulants as amphetamines and cocaine; nearly 13 million Americans use tranzuilizers, and barbiturates; heroin addicts in the United States are estimated to number 800,000 and may be actually much higher. The dangers of drug abuse are sharply illustrated by figures recently released to the Joint Congressional Committee on Atomic Energy: between March, 1972, and February, 1973, no fewer than 3,647 military and civilian employees with access to nuclear weapons were removed from their positions; 20 percent of them were removed because of drug problems. In addition, it was reported that 39 percent of all removals from the United States Army and 35.6 percent of all removals from the United States Navy during the same period were the result of drug abuse.[4] There is no doubt about it: alcohol and drug abuse are undermining America — mentally, morally, and physically. Now, we will review conditions of civilian drug abuse which frequently spill over into military installations.

GETTING HOOKED — WHAT IS ADDICTION?

Many experts agree that there are three steps leading to the condition commonly called drug addiction, though these particular steps may be characterized somewhat differently by different authorities. First, the individual must have developed what is labeled "tolerance" — an ability to tolerate increasing amounts of the drug so that the desired effect is obtained. One of the most unfortunate aspects of opiate addiction is that tolerance to the toxic, sedative, and analgesic effects of opiates can be nearly complete.[5]

Second, the individual must have become psychologically dependent on the drug. The Expert Committee on Drugs Liable to Produce Addiction of the World Health Organization has said that this step, often known as "drug habituation," includes four characteristics: a desire (but not a compulsion) to continue taking the drug for the sense of improved well-being that it engenders; little or no tendency to increase the dose; some degree of psychic dependence on the effect of the drug, but absence of physical dependence and, hence, no abstinence syndrome; and a detrimental effect, if any, primarily only on the individual using the drug.[6] Various other authorities have defined this second step as a "psychic craving" for the drug[7] and as a "compulsion to continue the drug and obtain it by almost any means."[8] Obviously there is considerable disagreement among experts.

The Expert Committee on Drugs has listed the characteristics of the third step, known as addiciton: an overpowering desire or need (compulsion) to continue taking the drug and to obtain it by any means; a tendency to increase the dose; a psychic and generally a physical dependence on the effects of the drug; and an effect detrimental to both the user of the drug and to society.

More recently, however, the Expert Committee on Drugs recommends that the term "drug dependence" be substituted for the terms "drug addiction" and "drug habituation." The committee now defines "drug dependence" as a "state arising from repeated administration of a drug on a periodic or continuous basis" and asserts that the characteristics of drug dependence will vary, depending on the drug involved.[9] Morris M. Rubin prefers to use the terms "psychological dependence" and "physiological dependence" to denote the second and third stages in the making of a drug addict.[10]

Alfred R. Lindesmith, a noted authority on drug addiction, has defined addiction as "that behavior which is distinguished primarily by an intense, conscious desire for the drug, and by a tendency to relapse, evidently caused by the persistence of attitudes established in the early stages of addiction."[11] He also writes that other correlated aspects are the dependence upon the drug as a twenty-four-hour-a-day necessity, the impulse to increase the dosage far beyond bodily need, and the definition of one's self as an addict.

Johannes Biberfeld, pioneer drug researcher, has posited the theory that opiate addiction involves both tolerance and craving for a drug, and that it is the phenomenon of craving which is uniquely human.[12] It is for this reason that many writers are unwilling to use the term "addiction" when referring to animals who have developed a tolerance for a drug in laboratory experiments. Nor will it help us much in our search for a definition of drug addiction if we do not proceed further after having learned the physiological changes in the human body that a particular drug can cause. While we must bear in mind the admonition of Harris Isbell not to regard physical dependence on drugs as wholly of "psychogenic origin,"[13] we would still feel justified after reading a volume on the physiological effects of drugs in asking, "Yes, but is this what we mean by drug addiction in human beings?"

What, then, is the key to narcotics addiction? It is not pleasure, according to one of the most widely accepted and cogent theories of addiction. It has been pointed out that if pleasure were the ultimate goal of the addict, marijuana would be the leading addictive drug; yet it is generally admitted that marijuana is nonaddictive, at least physiologically:

> *Marijuana seems infinitely superior to opium as a pleasure-producing agent; its pleasures do not fade as do those of opium with continued use; its psychological effects are described with enthusiasm and hyperbole; its pleasurable effects are not counterbalanced by the extensive evil social and physical consequences which ordinarily bedevil heroin addicts.*[14]

By this theory, the so-called craving for drugs is determined by negative reinforcement — that is, the relief and avoidance of discomfort and pain rather than by positive pleasure. The theory does not deny that an individual begins taking a particular drug because it gives him "positive pleasure," but it does refuse to label him an addict until he responds to withdrawal symptoms in a particular way — by craving the drug, both psychologically and physiologically. The real junkie, Jerome

H. Jaffe pointed out, seeks a drug with an overwhelming desire not to gain "pleasure" but to remain in a "normal state."[15]

Further, Lindesmith asserts that drug addiction, like almost all forms of behavior, is learned, for unless the person using a drug understands the reasons for his withdrawal symptoms, he will not become addicted to that particular drug. An individual who traces the pain associated with discontinuance of a drug to some other source than its discontinuance will not become dependent on the drug. This theory, while holding that the physiological or biological effects of drugs are indispensable preconditions of addiction, maintains that they are not sufficient preconditions:

> *Understanding withdrawal distress means to conceptualize it, to name and categorize it, to describe and grasp it intellectually through the use of linguistic symbols. Addiction is therefore a uniquely human form of behavior which differs from the superficially comparable responses of lower animals much as human cognitive capacities differ from those of lower forms.*[16]

Lindesmith adds that the habit-forming power of any drug is roughly dependent on the severity of withdrawal symptoms and not on the pleasures it produces. This theory would seem to be confirmed, at least to some extent, by the words of Thomas De Quincey in his famous work, *The Confessions of an English Opium-Eater:*

> *Lord Bacon conjectures that it may be as painful to be born as to die. That seems probable; and during the whole period of diminishing the opium, I had the torments of a man passing out of one mode of existence into another, and liable to the mixed or the alternative pains of birth and death.*[17]

Lindesmith's theory has its attractions, not the least of which is that it goes far toward drawing a clear line between those who use drugs and those who are addicted to them. He appears to define drug hunger solely in terms of avoidance of withdrawal symptoms. In addition, this theory helps us understand why the narcotics addict is a pathetic, helpless creature, caught in a habit over which he has very little, if any, control, for his craving "is not a rational assessment or choice of any sort, but basically an irrational compulsion arising from the repetition of a sequence of experiences in a process like those that lead to the psychologist's conditioned response."[18]

The "learning" theory of drug use may very well also apply to those substances which are not considered at least physically addictive. An individual, Howard S. Becker states, becomes a real marijuana user only if he or she learns to enjoy the sensations experienced.[19] Becker also argues that during this process, the individual develops a disposition or motivation to use marijuana which was not and could not have been present when he or she began use. Individuals who do not go through this threefold process while using the drug will not become marijuana addicts.

HEROIN AND ITS ADDICTS

> *You are fed up with everything for the moment. And like the rest of us you are a little hungry for a little hope. So you wait and worry. A fix of hope. A fix to forget. A fix to remember, to be sad, to be happy, to be, to be. So we wait for the trustworthy Cowboy to gallop in upon a white horse. Gallant white powder.*[20]

As long as heroin was mainly confined to the ghetto areas, white middle-class America did little to restrict its use and manufacture. When, however, heroin addiction became widespread among American servicemen in Vietnam and threatened to assume epidemic proportions among its own youth, white middle class America began desperately, and perhaps too late, to attempt to solve this tragic problem. The Bureau of Narcotics and Dangerous Drugs, a police enforcement agency of the Department of Justice, estimated that in March, 1972, there were nearly 300,000 heroin addicts in New York City alone.[21] The cost to America in terms of crimes committed in order to obtain money with which to purchase the drug is enormous. A depressing fact is that only 20 to 40 percent of addicts seek treatment voluntarily. The imprisonment of an addict costs nearly $8,000 a year and the rate of those released who return to prison is more than 90 percent.[22] Although comparable data was not available for the armed forces, most experts estimate similar waste of lives and money caused by drug abusers in the military.

How does the individual begin the heroin habit? Dan Waldorf has stated, in a study of 417 male heroin addicts, that 96 percent of those he interviewed said that they used heroin for the first time with one or more persons of the same sex and generally with friends.[23] The "communal" nature of narcotics addiction is well illustrated by the experience of one GI in Vietnam:

> *Before, we were all drinking beer. Only one man in our hootch was smoking* [*heroin*]. *Then one after another of us tried it and liked it; the whole place changed. Pretty soon we had rock music, posters, black lights, and we just sat around and smoked. We seemed to get along better.*[24]

As Norman E. Zinberg has pointed out, heroin is a social activity occurring mostly in small groups of friends.[25] The process of drug addiction presupposes membership in social groups and linguistic intercommunication.

In the abovementioned study, conducted by Waldorf and his associates, thirty-eight percent of the interviewees stated that they first tried heroin because they were curious about its effects, and thirty-six percent said that their friends were using it and they wanted to try it, too. However, it is difficult to give a general psychiatric diagnosis of the average heroin addict. As Marie Nyswander, a pioneer in methadone research, has written: "Addicts may be schizophrenic, obsessive-compulsive, hysteric, psychopathic, or have simple character disorders.[26]

Nyswander's observation is borne out by the remarks of addicts about the effects of heroin on them. Some of the reasons given by the addicts in Waldorf's study for getting high on the drug were:

> *It gave me peace of mind. I could get away from reality and forget my complexes. Straight, I felt I couldn't relate to people, and when I used heroin I could communicate better.*
>
> *I liked getting high. It was a good feeling. Heroin made me feel secure. I really felt protected. When I was high nothing could hurt me.*
>
> *Heroin makes you forget about your problems; makes you feel you know everything. You feel strong and healthy, not weak. You can work.*[27]

Vietnam veterans gave these reasons:

> *I just couldn't stand the pressure. You'd go into a ville and see a lot of your buddies blown up and it all didn't make any sense.*
> *I was nervous and frustrated and I wanted to see what it did for everybody.*
> *Somehow, it was easier to stand the heat of the stove and all the grease when you were high.*[28]

Heroin seems to give its victims, at first at least, an artificial peace. One cannot overlook the possibility that there are many more potential heroin addicts in our society than there are active ones. Perhaps each addict is the victim of deep, unexamined anxieties, is driven by them in a most painful way, and has only to find his way to the artificial peace that will allay them. It has been said that the heroin-user seeks fantasy instead of reality because his social condition or his personal condition, or both, are unbearable: "Taking heroin, at least on a regular basis, seems to be part of a desire to manipulate external reality."[29]

But this peace does not last long for the heroin addict; the addiction soon becomes a hell. The effects of initial experiences with heroin are short-lived, for the liver produces certain enzymes to detoxify the drug at a much faster rate than at first. Because of this process, the heroin user must, within a relatively short period of time, increase the amount and frequency of doses. In only a few months' time, the user must have the drug in order for his or her body to run "normally." If these people cannot obtain the drug, they suffer what are usually termed "withdrawal symptoms" — restlessness, irritability, shivering, muscular tremors, headaches, flushed skin, chills, dilated pupils, insomnia, abdominal cramps, excessive sweating, delirium, near convulsions, vomiting, diarrhea, dehydration, and weight loss.

Once the heroin user is "hooked," his or her entire life style revolves around the drug. As Waldorf describes it:

> *Soon all of life has an overwhelming purpose and focus. Life is simplified into a single, engrossing need that must be met before consideration can be given to any of the other physical or social needs that we think are ncesssary for any reasonable life. Nearly all activity is focused upon the day-to-day struggle to get the drug to satisfy a single need.*[30]

Dorothy Nelkin observed that the addict's concept of time is one of the most characteristic symptoms: it is adapted not to a schedule set by society by the day, or by the week, but "by the immediate demands of getting a fix to stave off withdrawal symptoms. Time is necessarily directed not to the future, but to present needs."[31] But the addict's sense of time is not all that is awry. Heroin users distort all aspects of their lives.

The lives of heroin addicts become inextricably intertwined with the lives of other users, for their addiction has created an ever-widening gulf between them and the "straight" world. Their condition makes it difficult for them often to communicate with others on what is commonly termed a rational basis. The heroin user's fellow addicts often reinforce his or her habit simply by their presence, for he or she receives gratification from association with them as well as from the drug itself. New friends are usually similar people and provide an opportunity to talk about and discuss drug experiences. Drug-pushers often spend hours at a time talking with their customers about drugs and their various effects, long after a sale has been made.

In addition to the above handicaps, the heroin user is often involved with the law. It seems impossible to obtain a fix on more than a day-to-day basis, and, thus, he or she risks arrest substantially more often than the typical marijuana user, who frequently builds up a supply of the drug to last a month or longer.[32] The heroin user is driven to criminal associations more often than most individuals in our society because the heroin supply is controlled almost entirely by organized crime. Branded a criminal, he or she begins to behave like one in a myriad of ways, and not simply through addiction. Rarely does the heroin addict voluntarily seek treatment and rehabilitation. When one does, "He is emotionally, socially, and physically exhausted; his money is gone, his family is alienated, his friends are in programs or in jail or dead."[33]

The high rate of heroin addiction in the United States should tell us something about our society as well as a number of things about our abuse of drugs. Nearly half of the adult population of the United States uses psychotropic drugs — sedatives, tranquilizers, and stimulants — at one time or another. No doubt that figure has increased recently. Father Patrick O'Connor, who has had long experience in dealing with student addicts, described the forces in our culture which make for drug addiction:

It is easy for us to say that their world, the drug-induced world, or maybe the mad scene in the discothèque, is unreal. But I think sometimes we have to sit back and look at our world: Our world of overindulgence in food and drink and pills; our world of divorce and alienation; our world of the double standard where the youngster will find a mother popping Demerol into her mouth and the father downing Anacin and Compoz; the uncle drinking two or three Martinis before dinner; our world with the twin values of "top dog" and "top dollar"; our world which was already addicted to drugs long before the current uproar about LSD or marijuana.[34]

ALCOHOLISM AND THE ALCOHOLIC

Perhaps there is as much variation in basic definitions of alcoholism as there is in the field of narcotics addiction. Alcoholism has been called by Thomas F.A. Plaut "a condition in which an individual has lost control over his alcohol intake in the sense that he is consistently unable to refrain from drinking or to stop drinking before getting intoxicated."[35] The term "loss of control" is taken to encompass two different phenomena: the inability to do without alcohol or to manage personal tensions without drinking, that is, the so-called "inability to abstain" and the inability to stop drinking after one starts. Drinking on and around military installations frequently starts as friendly "happy hours" but later becomes alcoholic nightmares.

The World Health Organization's Expert Committee on Mental Health has defined alcoholics as "those excessive drinkers whose dependence upon alcohol has attained such a degree that it shows a noticeable mental disturbance or an interference with bodily and mental health, their interpersonal relations, and their smooth social and economic functioning; or who show prodromal signs of such development."[36]

So complex is the phenomenon of alcoholism that Sydney Cahn regards it as a process whose defining characteristics must include a number of factors: the quantity of alcohol consumed; the rate of consumption; frequency of drinking periods; the effect of drunkenness upon the self and others; visibility to society's significant labeling agents (police, judiciary, employers, social peers, school personnel); the total social matrix of the person; the total problem syndrome of the drinking behavior; and the effectiveness of the formal and informal social controls.[37]

It is interesting commentary on our society that a study conducted by Eva and Richard Blum concluded that when patients were obviously from skid row or were apparent social misfits, doctors had no trouble diagnosing the individual's ailment as alcoholism.[38] But when the patient was well-dressed, had had no trouble with the law, and was not divorced, the physician was much more likely to label the illness at first as something other than alcoholism, even though alcoholism was later diagnosed. This is especially true in the military, where alcoholism is a serious problem among officers and noncommissioned officers.

Though there may be a range of disagreement among experts about what the ultimate defining characteristics of alcoholism are, few deny that the alcoholic, like the narcotics addict, must eventually pay a woeful price for the habit. Unfortunately, the more lasting effects of alcohol abuse on the body do not show up until the individual has continued heavy drinking for a number of years. Those systems which are most noticeably affected by prolonged alcoholic consumption are the digestive and the nervous systems. Digestive ailments may include gastritis, cirrhosis, and pancreas inflammation, while such nervous dysfunctions as neuritis, blackouts, hallucinations, extreme tremors, and chronic brain syndrome are far from uncommon. Sooner or later, the alcoholic ends up both a mental and a physical wreck.[39]

What are some of the characteristics of the alcoholic? It is estimated that five times as many men as women are victims of alcoholism in the United States and Canada, and alcoholism tends to occur more frequently among individuals aged thirty-five to fifty-five, though society today is finding itself burdened with an ever-increasing number of younger alcoholics. The alcoholic in the United States is more likely to come from Irish and Anglo-Saxon backgrounds than from Italian, Chinese, or Jewish backgrounds.[50]

David J. Pittman has advanced the theory that the large number of alcoholics in the United States is influenced by the fact that American cultural attitudes toward drinking contain social ambivalence — an oscillation between asceticism and hedonism — and thus restrict the development of stable attitudes toward drinking.[41]

Many authorities find that a number of alcoholics often have in common family situations that could lead to alcoholism when these children are grown. Studies suggest that the homes from which alcoholics come are characterized by authoritarianism, success worship, moralism, and overt parental rejection. A study of 502 children involving a thirty-year follow-up from the time these youngsters had been dismissed from a child guidance clinic showed that a significantly larger percentage of them became alcoholics than children who had not been to the clinic. Childhood histories of these patients showed that antisocial behavior on the part of their fathers was particularly frequent.[42]

Many psychoanalysts characterize the typical mother of an alcoholic as "an overindulgent and pampering person who often shields her son from a harsh, severe and inconsistent father." They further maintain that the child continues to shield himself from reality through alcohol and thus can gratify his infantile impulses. The alcoholic, it is pointed out, needs support and affection without effort. This school of thought posits the theory that alcoholism is an early fixation at the oral state of psychosexual development. Louis A. Faillace and Robert F. Ward concluded that when the individual learns that he or she can obtain this type of support and affection and avoid what is often taken to be a harsh reality through the use of alcohol, he or she comes to drink more and more and thus becomes physiologically addicted.[43]

Alcoholics, at least to a greater extent than a number of other individuals, suffer from an unusual amount of stress and a good deal of deprivation. They have difficulty in coping with both their frustrations and their impulses. Further, it is a problem for them to develop close and meaningful interpersonal relations. In many ways, alcohol is a "social lubricant" — it helps the alcoholic relax and be friendly. Donald A. Overton offers the theory that alcoholics may drink to obtain access to behavioral patterns or emotional reactions which have become conditioned to the presence of alcohol, rather than to obtain any of the intrinsic effects of alcohol.[44]

Few experts would agree that there is any such a thing as the alcoholic personality, but Ronald J. Catanzaro believes that at least thirteen characteristics, or combinations of them, frequently appear in the alcoholic: (1) a high level of anxiety in interpersonal relations; (2) emotional immaturity; (3) ambivalence toward authority; (4) low frustration tolerance; (5) grandiosity; (6) low self-esteem; (7) feelings of isolation; (8) perfectionism; (9) guilt; (10) compulsiveness; (11) angry overdependency; (12) sex role confusion; and (13) inability to express angry feelings adequately.[45]

Charles E. Goshen puts it another way:

> *We can say that the typical alcoholic exhibits characteristics that have some similarity to both the neurotic and the sociopath. He is like the former in that he seeks to avoid responsibility and tries to justify his behavior; he is like the latter in that he also defies social pressures. The fact that both sets of characteristics exist in the alcoholic is largely responsible for the fact that he is regarded by society as sometimes "sick" and at other times as "bad."*[46]

Can we predict who will and who will not become alcoholics? Some authorities believe that an individual who (1) responds to beverage alcohol in a certain way, often physiologically determined, by experiencing intense relief and relaxation; and who (2) has certain personality characteristics, such as difficulty in dealing with frustration, depression, and anxiety; and, who (3) is a member of a culture in which there is both pressure to drink and culturally induced guilt and confusion regarding what kinds of drinking behavior are appropriate, is more likely to become an alcoholic than will most other persons. Other experts state that the form of alcoholism which is indicated by the most commonly recognized symptoms can be predicted by indicants of anxiety neurosis, employment difficulties, disruptive interpersonal relations, and acting out rebellion.[47]

DRUGS, YOUTH, AND SOCIETY

As we study the GI drug user today, we find more often than not an individual at war with family, school, religion, and government — in conflict, directly or indirectly, with most of the major institutions of society. It would be comforting to many of us to be able to say that the young are pampered, that they have picked their war with society with little or no reason. Comforting, but simply not true. As James C. Bennett has reminded us, "Abnormal behavior often signifies that the individual is out of touch with adult society. It can also mean that adult society is out of touch with the individual."[48] And Bennett cautions us always to remember that it is the job of those trained in interpersonal relations to keep both the society and the individual in touch.

Many adolescents maintain that drug use enables them to establish contact more easily with others, to break down the barriers, rather than fulfilling the traditional goal of producing and achieving with others. However, drugs allow the user to relate to others without being really close.[49] The adolescent user views the values of society as inimical, dehumanized, and hypocritical. An unusually large number of adolescent drug-users come from cultured, affluent homes, and from parents who are well-educated and socially prominent. In fact, these youngsters complain that so much of their parents' time is taken up with the pursuit of material prosperity and social prominence that little time is left for their emotional needs.

Many such adolescents view their fathers as passive, remote, compliant, uninvolved, disinterested, and ineffectual, and their mothers as dominant, assertive, active, intrusive, and overprotective. Often the mother in such a family is seductive with her children and encourages a symbiotic relationship. This description varies little from the picture of the male alcoholic's mother so often painted by psychiatrists. The upper-middle-class adolescent drug-user also often complains that his parents employ a double standard of morality that corrupts their entire lives. As one youthful marijuana user tells it: "Sure I use pot. It feels good. My God, my dad really thinks that makes me an addict. He's a scared man. When I tell him he smokes, drinks, and uses tranks [tranquilizers], he says, 'Two wrongs don't make a right.'"[50] This is a common complaint of young GIs being punished by older officers and noncommissioned officers.

Among high school drug-users, according to one recent important study, five times as many boys as girls are users. In comparison with nonusers, they were generally less religious, more involved in political organizations and underground groups, had poorer grades, were less happy with the manner in which they had been reared, were likely to come from broken homes or less close ones, and were more likely to have parents, siblings, and close friends who used or had used drugs.[51] Enlisted men and women recently out of high school conform to these general characteristics.

In a pioneering study of the social attitudes of college drug-users, Edward A. Suchman reported what he called the "hang-loose ethic" among such individuals. This ethic attempts to build another value system, one divorced from the establishment set of values, whose goal is freedom from conformity, and whose search is for new experiences. Suchman listed nine characteristics of college students who regularly smoked marijuana: (1) dissatisfaction with the education they were currently receiving; (2) a belief that students should have a more meaningful role in educational decision-making; (3) opposition to the Vietnam war and the draft; (4) approval of premarital sex; (5) a belief that it would be acceptable to evade the law if one did not actually break it; (6) a feeling that there was a serious communication gap between parents and students; (7) an expectation of deriving satisfaction in future life from leisure activities; (8) participation in 'happenings" and large protests; and, (9) the reading of underground newspapers.[52] The war and draft are over but the other characteristics remain.

Other studies of college students who use drugs regularly show that they are a good deal less likely to be or to become affiliated in any way with organized religion. On the contrary, they are much more likely than nonusers to profess atheism or agnosticism.[53] It has been suggested that cannabis intoxication may be a form of religious experience, however unorthodox, for the adolescent to whom organized religion no longer offers a spiritual challenge and relevance. One writer sums up this attraction to drugs by saying:

> *Man has a very deep craving to go beyond the narrow bonds of a mortal existence. Modern man escapes this permanent and fundamental drive no more than his ancestors did. When the rising generation uses drugs to quench its thirst for evasion, it is reverting to an old mystical practice in which the most primitive tribes have always indulged.*[54]

It is not altogether surprising that our schools, at the elementary, secondary, and college levels, have registered such an alarming rate of drug addiction. A very large number of students believe that they have no control over their academic lives. Some point out that our schools fail seriously to provide relevant education, and many experts believe that this is one of the major reasons why ever-increasing numbers of adolescents are turning to immediate and certain rewards in the form of drugs. Students should be considered *subjects,* not objects. Schools cannot combat drug addiction until "self-motivation replaces coercion; enjoyment of learning replaces grades; and stimulators and resources replace disciplinarians."[55]

It is all too often the very bright student who will not conform, who will not play a "subordinate" role in school, and who is likely to become the drug user. As his or her personal needs for self-esteem, identity, and maturity are frustrated by a paternalistic educational organization (one that is experiencing an identity crisis of its own), he is more and more attracted to an escape that is all too easy.[56]

Educational institutions at all levels — including military programs — have come close to refusing any responsibility at all for the emotional health of their students. They must take up the task of helping alleviate the emotional problems of drug users, who, as a group, have approximately twice the incidence of mental disturbance of nonusers. The emotional disturbances of their students, however, will go largely without treatment until the schools attempt to restore some meaningful form of communication with the young. Rigid authoritarian classrooms which characterize military schools are the least likely to provide for two-way communication between teacher and student.

What can never be stressed strongly enough is that the epidemic of drug use among our GIs is not some isolated phenomenon whose prevention and cure can be considered apart from other social conditions. As Gabriel G. Nohans has stated, "Young people emerging from a protected childhood are appalled to discover the many unresolved contradictions of modern American society in the areas of race relations, distribution of wealth, health care, urban congestion, environmental pollution, and foreign policy."[57]

A number of experts agree that the chief predependency motive of the drug user is an inability to cope with his total environment. Drugs may offer a way for the young user to adjust to an environment which would be unhealthy for any human being. "Drug abuse," as Richard Brotman remarks, "may be one of the expressions of the individuals's isolation in society, as a byproduct of the stress of technological advance."[58] Are we living in a society that is "sick," as a number of young drug users have charged? The fact that so many of our youngsters perceive it to be sick, while it actively persecutes them for what they consider to be harmless behavior, is a paramount issue. Some writers go so far as to assert that marijuana usage is a socio-political issue that is a polarizing force in society.

What many decision-makers in the armed forces may have failed to realize is that for a large number of GIs, the lines to the past have all but snapped, and it is very difficult to place the responsibility for this tragedy on their shoulders: "This failure in communication and disruption in an orderly transmission of values occurred because most of the older generation failed to realize the seriousness and extent of the [GI's] distaste for the goals that had satisfied previous generations."[59]

The older generation has not communicated with its youngsters for a long time, perhaps more from inability than from unwillingness. Marlin H. Dearden may have given us a key to the future:

> *Until we are willing to listen to students and to each other, and are then willing to explore together those areas of personal concern as they exist in individuals and organizational and administrative operation, our efforts in drug education, and for that matter, education, will be less effective than they could be. In effect, we will be bandaging a patient who needs an operation.*[60]

In conclusion, it must be pointed out that one of the characteristics which alcoholics, heroin addicts, and young drug abusers share is the inability to communicate with the so-called "straight" world. It is not enough to say that the burden of responsibility for relating is on those who deviate from establishment behavior and norms. Responsibility in such situations is always a two-way street. Perhaps one major priority for the future in dealing with alcohol and drug-related abuses in the military should be the pursuit of techniques that facilitate dialogue between those who uphold society's values and those who, for whatever reasons, have challenged these values, at least implicitly, by taking drugs. Only through such a dialogue can we objectively reexamine our own values to see whether our view of reality is one that still entirely coincides with the conditions for human behavior. In the meantime, the problems associated with drug and alcohol abuse continue to plague our armed forces.

The Vietnam Connection

Colonel Olen D. Thornton, Executive Officer

2nd Basic Combat Training Brigade
Fort Leonard Wood, Missouri

VIETNAM

In the mid-1960s, reports of an epidemic increase in the use of mind-altering drugs by U.S. military personnel began trickling in from Southeast Asia. Initial reports reflected an increasing use of marijuana by American troops in Vietnam. Despite government attempts to deal with the problem, drug use became a way of life for a growing number of American servicemen. In mid-1968, the Army cracked down on both marijuana-smoking GIs and the Vietnamese suppliers. A full-fledged assault to stamp it out got under way. Radio and TV spots proclaimed the evils of marijuana and indicated not only that a smoker could damage his own brain and become psychotic, but that his apathy toward the rest of the world could result in injury to men dependent on him.

About the same time, soldiers from Thailand entered South Vietnam to reinforce the American troops. These reinforcements brought with them "red-rock" heroin, an inferior grade of heroin (3 to 4 percent pure) heavily adulterated with strychnine, caffeine, and various inactive fillers. This was the first documented instance of heroin being made available to U.S. servicemen. After the Thai troops were withdrawn, heroin was once again a relatively scarce commodity. The Army had applied sufficient short-term pressure against marijuana use and trafficking during this period (in one week of the campaign, 1,000 arrests for possession were made) to justify press releases expressing official optimism that the "drug problem was under control." Consequently, enforcement efforts against marijuana somewhat abated.

A similar pattern occurred between May and July, 1970. Enforcement efforts against marijuana were again intensified. Concurrent with this pressure on marijuana use, vials of 96 percent pure heroin began appearing in Saigon. By the end of 1970, heroin was readily available throughout South Vietnam. These two occurrences — the Army's crackdown on marijuana and the accessibility of heroin — had an effect on drug-use patterns totally unanticipated by the military. The result was that GIs who had been smoking only grass turned to snorting and smoking heroin, which was initially passed off to them as nonaddicting cocaine. They reasoned that the substance itself, heroin, and the smoking of it were more easily concealed from prying eyes and noses than marijuana. Increased Army pressure against heroin decreased the supply, and users were forced to employ a more efficient way of getting the drug into their systems. Those who had previously been snorting or smoking heroin graduated to mainlining.

The experience for thousands of American troops was roughly the same. Young soldiers began snorting and smoking the pure Vietnamese heroin in the tragically mistaken impression that they could only become addicted by shooting it into a vein.

Domestic concern over an influx of young men with extensive drug experience and/or physical addiction rose to a point of near panic. Escalating drug use in the United States and abroad led President Richard M. Nixon, on June 17, 1971, to direct urgent and immediate attention to a national counteroffensive against drug abuse. In turn, the secretary of defense directed the service secretaries to develop drug-abuse prevention and control programs.

SURVEYS

Several official and unofficial surveys have been conducted in an effort to determine the dimension of the Army's drug problem. The illicit use of drugs became so widespread in Vietnam that it was almost impossible for the military enforcers to identify, apprehend, and discipline any but the most obvious offenders. Estimates of the number of servicemen who became physically addicted to heroin or other opiates in Vietnam ranged from what I consider a conservative figure released by the Department of Defense (4.5 to 5 percent) to the liberal estimate (25 percent) of Bentel and Smith. The results of a survey conducted by Human Resources Research Organization (HumRRO) during the period September, 1970, to September, 1971, indicated that substantial numbers of enlisted men used nontherapeutic drugs. The usage rates varied form 29.9 percent for marijuana to 11.7 percent for narcotic drugs. Drug usage rates in the Army were sufficiently higher than the rates in the other services. Army enlisted personnel reported daily usage rates of approximately 3 to 4 percent for illegal drug stimulants, depressants, and hallucinogenic drugs other than marijuana. The estimated overall Army daily narcotic usage rate was approximately 5 percent, ranging from an estimated 9.2 percent in Vietnam to a low of 1.6 percent in Europe. The reported daily narcotic usage rate for Army men in the continental United States was about 5 percent. According to the April 24, 1973, issue of the *Washington Post,* a Pentagon study released by Dr. Richard S. Wilbur, assistant secretary of defense for health and environment, indicated that a follow-up study of returning war veterans showed that 35 percent of the Army enlisted men in South Vietnam tried heroin and that 20 percent of the GIs surveyed reported becoming dependent on the drug at one time or another during their tours of duty. These two studies illustrate the differences of surveys and point out the difficulty in establishing "official" figures regarding drug use and addiction.

AFTER THE VIETNAM WAR

Troop withdrawals from Vietnam and reduction of the Army force structure reduced the actual numbers of soldiers involved in drug abuse, but in terms of percentages, there appears to be no significant reduction. What happened, of course, was that the problem moved with the population and the availability of drugs. In the meantime, the problem had significantly increased in Europe. There are no recent official

Department of the Army surveys regarding the extent of drug abuse. Unofficial reports, however, from various Army agencies and from sources such as military police, criminal-investigation divisions, and counselors indicate little change in the usage rate, but they do indicate changes in use. In Europe, for example, the use of heroin is down. Use is somewhat controlled by availability. Use of stimulant amphetamines ("speed") has increased in the past few months, as well as the use of depressants, barbiturates, and alcohol. Of significant concern is the increased multiuse of barbiturates *and* alcohol. There is little or no use of marijuana, but the use of the potent hashish has increased significantly, primarily because of its availability.

Possibly the greatest failure in the Army's war on drugs has been the inability to shut off the flow to military bases. Illicit drugs can be bought on and near most U.S. and overseas military installations. The people who are sometimes identified as pushers — base employees, maids, bartenders, prostitutes, taxi drivers, and even kids from nearby towns or villages — still find a ready market among GIs.

Although there is no official agreement or empirically substantiated documentation regarding the overall Army rate of drug abuse, there is widespread agreement that it indeed remains a significant problem in all major Army commands.

CAUSE FACTORS

Living conditions, easy access to drugs, alienation, boredom, escape, rebellion, and pushers have all been blamed for the problem. While any of these factors may be a contributing factor, no single cause or set of conditions clearly leads to drug dependency; it occurs in all social classes or contexts. The key factor is the abuser himself. Thousands of troops are exposed to drugs each day, and, while comparatively few turn to a life of drugs, many young soldiers are joining the ranks of the abusers.

Many will say that the Army's drug problem was "caused by Vietnam." Perhaps that statement is justified, but let us take a deeper look at some of the causes. Many studies indicate that sizeable numbers of servicemen who used drugs in Vietnam had exposure to or experience with illegal drugs before entering the service. Normally, drug abusers in Vietnam were not considered highly deviant men. They were soldiers who were alone, away from home, often resentful — and ignorant. One soldier reported, "I got into scag four hours after I got to Nam. We were all smoking grass and this guy gave me some scag, put it in the end of my joint. Man, I just wanted to get out of Nam and scag took me out — for a while, at least."

Stepping off the plane into the smothering heat of Vietnam, the young soldier was confronted with a mixture of exotic sights, sounds, and odors. He arrived with visions of John Wayne heroism dancing in his head but soon found himself sinking into the reality of the war zone. From the pusher's point of view, he knew not only when many soldiers were psychologically ready but also that they had a known payday — and that pay was often pure spending money, since many other expenses were deducted or prepaid by the Army.

Participation in the war was a particularly painful, lonely, boring, and confusing experience for many GIs. Many were either unclear about the goals of the U.S. involvement or violently disagreed with the stated domino theory and the save-the-world-from-Godless-communism line. Many of the jobs were seen as meaningless and routine; the realization that many of the people back home did not support the war was an ever-present weight; the combat soldier lived under the constant threat of imminent death or injury. These pains were intensified by the less-than-cordial relationships between many of the U.S. troops and their South Vietnamese counterparts, who viewed each other with distrust and contempt. By the Americans, the South Vietnamese troops were often seen as cowardly and unreliable and considered potential or actual enemies.

It is now common to find groups of young people whose central interest and acitvity is drug use. The Army environment is no exception. Social pressure has long been a powerful force in determining behavior, especially with regard to using alcohol or other drugs. Certainly, to be "in," one has to participate in the activities of the members of his peer group. He is in a circle, and they are passing the joint around, and he wants to be part of the in group. So he "takes a hit," and after awhile is hooked.

Many of the fighting men in Vietnam depended on drugs to alter and enhance reality. They used drugs to manipulate time — time that moved them closer to their DEROS (Date of Expected Return from Overseas).

Increasing drug use toward the date of one's DEROS was a common occurrence, as it seemed to ease the paranoia of the short-timer.

Poverty, racial friction, urban decay, campus unrest, family-marriage breakdown, the Vietnam war — all are factors that influence our youth. Many of these same pressures follow the soldier into the service, and, unfortunately, the realities of military life often tend to compound these pressures and create new ones. The demanding nature of many day-to-day activities tends to exacerbate an already tense situation. Because reality is sometimes very painful, drugs may seem to be the best, or at least one, means of relieving this pressure. Closely akin to the social pressures and tensions is the impact of boredom. The young drug user typically fails to feel any sense of self-importance. The daily routine around the barracks presents no challenges, no sense of accomplishment, no pride. The young drug abuser is often merely looking for a challenge, some excitement. And when this excitement — drugs — is accompanied by an aura of glamor, and disapproval from the establishment, the thrill is doubled.

The causes and motivations underlying drug-abuse behavior are extremely difficult to ascertain. Those experienced in this area stress the diversity and complexity of such factors. Although many drug users have been significantly influenced by peer pressure, the decision to abuse drugs is essentially an individual one. Although such decisions are sometimes made in a social vacuum, they must be consummated by the individual himself.

REHABILITATION PROGRAM

Drug abuse has a particularly important consequence for the armed forces. Those in the military service have a special dependence on each other. The lives of those in a unit may depend on the alertness of one man assigned to a specific task. No commander can trust the fate of his unit, ship, or plane to a man who may be under the influence of drugs. The drug abuser in military service is a security risk — for example, he can be blackmailed by threat of exposure. He can also be led to sell or give away classified information to support an expensive drug habit. Additionally, while under the influence of drugs or alcohol, he may overlook or ignore proper security measures.

In accordance with the President's message of June, 1971, measures were established to attempt to identify, treat, and rehabilitate all service members dependent on drugs. Consistent with legal constraints, soldiers being separated from the service for any reason who were identified as drug-dependent were to receive a minimum of thirty days of treatment before discharge. Treatment was to be either in military facilities or in Veterans Administration (VA) centers. Those remaining in service were to be provided treatment and rehabilitation in military facilities; if they needed extensive treatment, however, they were to be discharged from the service after transfer to a VA hospital. The Army's initial major effort was to identify Army members departing Vietnam who were using or were dependent on narcotics. Personnel so identified were to undergo five to seven days of detoxification treatment before their return to the United States.

The desired method of identifying individuals who abuse alcohol or other drugs is by their asking for assistance. The Department of the Army Exemption Policy stated that a soldier who volunteered for treatment would not be subject to any disciplinary action under the Uniform Code of Military Justice (UCMJ) for his past use or incidental possession of drugs. Further, if he could be effectively treated and rehabilitated within the service, any discharge resulting solely from his past use or incidental possession of drugs would not be under dishonorable conditions.

Some individuals with drug-related problems simply would not volunteer for treatment. There was a universal lack of trust of the initial "amnesty program." Some of this lack of trust was justified; medical records were not always kept confidential, and after the one-shot rehabilitation, they were sent back to their units and were formally labeled as deviants and were treated as such by those in authority.

The effectiveness of this policy has improved since its initial implementation. The program is now called the exemption policy. Soldiers have learned that it provides exemption from punishment under the UCMJ or from a less-than-honorable discharge based solely on drug use up to the time a soldier asks for help under a voluntary treatment program.

Unit commanders are responsible for making it clear to the soldier that before he makes any disclosure about his drug abuse, other administrative action may include having his MOS (military occupational specialty), hazardous-duty orders, or security clearance temporarily suspended or his name removed from access lists for classified information while an assessment is made of whether the GI's drug problem hurts his ability to perform his duties; being discharged for drug use with an honorable or general discharge if the soldier is unable to respond adequately in a reasonable period (which requires him to go into a longer term civil program); or being punished under the UCMJ, or given a less-than-honorable discharge for his offenses — such as pushing drugs or stealing to support his habit — that are detected independently.

The exemption is not granted by anyone — it is automatic. It does not apply, however, to an offense committed after a soldier has asked for help. If he gets into drugs again and is caught breaking the law, he is liable for his new crimes.

The Army also recognizes the danger of backsliding. But if a soldier is performing effectively and is really trying to kick the habit, he will get the benefit of the doubt. There is some practical limit, of course, to the number of times a soldier can be allowed a relapse without raising serious questions about his ability to perform his duties. That limit is determined by the individual's commander. Although a commander cannot deny exemption, he has the option to determine how long help should be provided and to recommend an honorable or general discharge before sending the individual to the VA for help.

Methods of involuntary identification are command observation and apprehension of offenders and biochemical analyses of urine specimens to determine the presence of amphetamines, barbiturates, or opiates. Persons reported with confirmed positive results are provided treatment, detoxified, and counseled, as appropriate. Evidence developed by urinalysis administered for the purpose of identifying drug users may not be used in any disciplinary action under the UCMJ or as a basis for supporting, in whole or part, an administrative discharge under other than honorable conditions.

Within the areas of identification, some difficulties were experienced. The screening test requires that an individual submit a urine test for detection of drugs. An early difficulty was associated with a need to assure that a given urine specimen was submitted by a given individual. It was reported that servicemen were substituting everything from beer to their grandmother's urine in the tests. During the period of maximum testing in Vietnam, the requirement was initiated that the urine specimen be provided under conditions of direct observation by reliable personnel. This requirement caused a considerable delay in the collection process. The requirement today, however, that each individual submit a urine specimen — on the average of two times a year — can be met without difficulty.

The most dramatic part of the Army's drug-abuse program is the urinalysis-screening system, which identifies the drug user. But once you have identified him, what do you do with him? The first answer is "Separate the man from his drug." The medical term is "detoxification" — in lay language, "withdrawal," or to put it more bluntly, "dry him out." There is more to detoxification than drying him out, more to withdrawal than separating the man from his drug. In fact, *detoxification is the beginning of treatment.* Detoxification involves terminating physical dependence on drugs, or acute intoxication, and treating the symptoms that result. How a soldier is detoxified strongly influences his participation in later treatment and rehabilitation. A medical facility should be used when available and judged appropriate by the medical staff. The time spent in detoxification varies with the individual, his degree of drug dependency, and the drug or combination of drugs involved. No individual should be released from in-patient status until his withdrawal is complete and a medical assessment has been made. Then the third stage, *treatment,* begins.

Consistent with applicable statutes, a soldier found to be dependent on any drug will receive a minimum of thirty days of treatment before being separated from the service. This policy also applies to any soldier identified as an opiate abuser, regardless of the degree of his involvement. The thirty-day period includes time spent in detoxification, medical evaluation, in-patient hospital care, rehabilitation facilities, out-patient programs, stockades, and VA facilities before discharge from the service.

Separatees who are drug-dependent or opiate abusers may elect to extend their services to undergo a full thirty days of treatment in military facilities. Those who do not so elect will be transferred to a VA hospital before discharge to complete at least thirty days, or as much of the thirty days of required treatment as statutes permit. Soldiers transferred to a VA hospital will have fifteen days remaining before their discharge whenever possible.

Separatees identified as drug users are advised of the opportunity for volunteering for continuing treatment in military facilities and of the availability of VA treatment. When such treatment options are declined, the fact of the individual's preference will be made a matter of record.

The percentage of those who have volunteered for treatment has been relatively low. The most-often-stated reason is that the extended time is "bad time" since it extends the soldier's time in the service.

With a group as angry, frustrated, and difficult as the drug-using soldiers leaving Vietnam, involving them in treatment was a complex and tough job. In detoxification and treatment, we found that our stereotypes of expected behavior did not hold true. In part this was due to a lack of organizational experience in dealing with drug users and in part to the drug users' experience in manipulating the establishment. To put it another way, drug abusers in detoxification did not respond as we expected. We are beginning to find ways to create a working alliance with our drug-dependent patients. It is a classical case of *how* you do something being more important than *what* you do.

Medical care is necessary. Far more crucial is to cut down on the medical mystique of "detoxification and treatment." The most beneficial thing we can do for a man in detoxification is maintain and bolster his links to his unit. Only rapid rehabilitation — return to unit — can keep the man thinking of himself as a *soldier* and not a *patient*. And that is the final goal of detoxification and initial treatment.

It is the Department of the Army policy to rehabilitate members who evidence a capacity to undergo such rehabilitation. The objective of the Army rehabilitation programs is to return the soldier to full effective duty with short-term rehabilitation efforts or to provide continuity of care for separatees. Entry into a rehabilitation program may be voluntary or involuntary — if, through evaluation by the commander, chaplain, medical officer, and legal officer, it is determined that an individual who has been involved in an incident of drug abuse or who has been convicted by civil or military court for drug abuse has rehabilitative potential.

The willingness of the respondent to volunteer for drug treatment and rehabilitation is a vital factor in the success of any program. Headquarters, USAREUR (U.S. Army Europe) reported in mid-year 1971 that only about 10 percent of those in treatment for drug-related problems were under the then so-called amnesty policy. Several administrative shortcomings have been eliminated since then, and the percentage has increased.

Most medical authorities agree that drug abuse is a manifestation of a deep-rooted personality disorder and that successful rehabilitation usually involves a reorientation of the character of the abuser. This is no easy task, particularly since it is normally a lengthy process. Some soldiers will need sixty days or more of rehabilitation. Because of the relapsing nature of the condition, one or even more temporary reversions to drug abuse should not be considered sufficient grounds to adjudge rehabilitation a failure. The soldier's attitude, work efficiency, and potential must receive primary consideration. When a drug abuser is found to be unwilling or unable to respond adequately to genuine rehabilitative efforts, he is administratively separated from the service after transfer to a VA hospital.

Supplementary rehabilitative services are usually provided in a halfway house or rap center. Halfway-house facilities provide a structured environment for the individual who does not require in-patient care but who is not yet ready to function on his own. Such facilities provide for a man to "live in" either full-time for a short while or part-time while performing duty in his unit. The soldier should not be separated from his unit and the realities of military life for extended periods. Although medical personnel will supervise the professional and paraprofessional aspects of rehabilitation, the halfway-house program is a command responsibility. Rap centers play an important role in the out-patient rehabilitation program; many soldiers do not need contact with a halfway house, and other, more resistant soldiers will respond better to a less structured program.

Although the overwhelming majority of drug abusers within the Army are not dedicated career soldiers, the Army recognizes the obligation to do everything within its capabilities and resources to help those who have become involved with drugs. Medical personnel play a major role in rehabilitation through psychiatric treatment, counseling, and hospitalization if required.

PREVENTION

Although the Army has planned for the rehabilitation of soldiers who are already in the grip of drugs, prevention is the answer for the future.

Anything less than a successful program intended to prevent and treat dependence on drugs and alcohol will have an adverse effect on the Army — both internally in terms of morale and discipline and externally in terms of public support of the Army.

Prevention includes leadership, education, and law enforcement. The basic factors that lead to alcohol and drug abuse — lack of responsibility, boredom, lack of meaningful activity, ineffective channels of communication, stress, and frustration — must be overcome through sound leadership. Leaders at every level must meet the problem head on. The fight must be waged and won at the company level: that is where the soldier lives and works. How can this be done? Some suggestions provided by Brigadier General Robert G. Gard, Jr., Director of Discipline and Drug Policies of the Army, include good man-to-man communications and programs that show someone cares, to help put down the feeling of alienation; produce meals worthy of the mess hall; provide recreational facilities that "grab" the troops; plan and execute an athletic program in which every unit member can participate; and make sure the man has an alternative way to get high. The Army needs quality leadership today more than ever before. I firmly believe that we are moving toward an enlightened leadership that practices firmness but not harshness, generosity but not selfishness, and pride but not egotism.

Getting facts across to troops is traditional in Army education. Facts include information about drugs, the reasons behind drug abuse, and ways of coping with the problems that result. Past programs assumed that if men knew the possible physical and legal consequences of drug-taking they would not abuse drugs. This assumption was wrong; knowing the facts is only a necessary first step. Influencing attitudes and behavior is more difficult and more important. We have learned to base our programs on honesty and objectivity and we have developed innovative approaches that have credibility with the target audience. Unconventional methods, such as small-group sessions and theatrical productions, are now being used. Officers and noncommissioned officers are being trained to lead small-group discussions. Natural leaders, who strongly influence behavior and attitudes of their associates, are being identified and utilized.

The principal objectives of law enforcement are to eliminate the sources of illegal drugs and to apprehend traffickers. Prevention efforts are being improved and are being coordinated with civilian employees, dependents, civilian communities, and civilian law-enforcement officials.

While the Army stresses empathy and compassion in dealing with individuals, we must also stress that the American taxpayer does not want his tax dollar creating a laissez-faire environment within the Army. Discipline is a vital part of any military organization. Although many factors are involved, there is some correlation between the lack of discipline within an organization and the drug problems. Our military force must be dedicated to high standards of personal conduct and dedicated service. We must create an environment that will appeal to the soldier, where he is treated like a mature, responsible individual.

ALL-VOLUNTEER ARMY

There are some who contend that the all-volunteer Army will become an Army drawn from the less privileged in America — peacetime mercenaries who decide on financial grounds that the military has more to offer than years of picking tobacco, sorting potatoes, washing cars, waiting on tables, or looking for better jobs that are not there. Several studies have been conducted to determine the type of young man who is now willing to join the Army. The profile developed from these studies indicates that he is a young man about nineteen years of age, with a fine upbringing, and reasonable expectations. He is a high-school graduate, raised in a blue-collar or lower-middle-class environment. He likes his home but feels that it is time to leave and become his own man, and yet he wants the security offered by an institution. He is a little frightened of basic training, but he will try his best. At the end of basic training, he is at the peak of satisfaction, and he is convinced that this is the best thing that ever happened to him. The typical female recruit has similar characteristics.

The young recruit has several expectations of the Army. If he or she is treated as a mature, responsible individual and provided with job satisfaction, he or she is not likely to become a drug or alcohol abuser.

Rehabilitation: One Man's Opinion

Captain Thomas S. Eisenhart, Drug/Alcohol Abuse Officer

Altus Air Force Base, Oklahoma

On June 11, 1971, President Richard M. Nixon declared that drug abuse was the public's number-one enemy. On the same day he sent a memorandum to the Department of Defense directing all military services to conduct programs for the identification and treatment of drug abusers. The intent was to avoid returning members to civilian status in a durg-dependent state. This marked the beginning of organized efforts within the Air Force to deal with drug abuse at the installation or base level. People were selected, trained (to varying degrees), and put to work in an effort to defeat this public enemy. The following pages cover in part the energy expended and report the state of affairs at one particular Air Force base at the time of writing.

In this report, I will examine the rehabilitation groups, paying particular attention to a basic dissimilarity between groups within the military and those outside the military services. This distinction is, in my opinion, the fact that civilians generally enter rehabilitation at their own request, whereas military personnel may be placed in rehabilitation against their wishes. Thus we will see the distinction between voluntary groups and forced groups. Some evidence from my own experience will be presented to support the concepts presented.

Before considering the rehabilitation groups (the raw material), I will scrutinize rehabilitation itself, as a process. At that point I will explain my personal bias and share with you, the reader, rehabilitation as I see it. Again, I will draw supportive evidence from personal research experiences. I will also discuss my concept of rehabilitation, since it offers some degree of change from the rehabilitation experience. I will explore the need for a change-measuring device, the device itself, and the effectiveness of this particular instrument. My personal thoughts and comments follow.

AIR FORCE POLICY

The current Air Force policy and guidance concerning rehabilitation is contained in Air Force Regulation 30-19, "Illegal or Improper Use of Drugs" (Section G), and Air Force Regulation 30-23, "Alcohol Abuse Control and Rehabilitation" (Section E). For the benefit of those readers who are not familiar with the base or installation rehabilitative efforts within the Air Force, I will briefly summarize the regulations.

Air Force Regulation 30-23 is very broad and general in the guidance provided. Thus, at the base where this research was conducted, it was decided to pattern the Alcohol Rehabilitation Program after the Drug Abuse Rehabilitation Program. AFR 30-23 does, however, state that initial treatment efforts will be provided whether the individual is a volunteer or a nonvolunteer. Similarly, Air Force Regulation 30-19, paragraph 22, states, "In all cases when a member is identified, the unit commander will document the date and means of identification and forward the information to the Chief of Social Actions by official letter. This letter constitutes official entry of a member into rehabilitation." Thus both programs contain an avenue for entrance of persons who, for whatever reasons, do not choose to engage in personal treatment of this nature. Herein lies the seed of what I call "forced groups," and I shall deal extensively with the growth of these in the next section. For the present, let us take note of its existence and pursue the concept of rehabilitation as outlined in Air Force directives.

AFR 30-19 outlines rehabilitation as a five-step, or five-phase program, the objective of which is to return the individual to full-duty status or to assist him in his transition to civilian life. These are the steps or phases:

Phase I. Identification
Phase II. Detoxification
Phase III. Medical evaluation and treatment
Phase IV. Behavioral reorientation
Phase V. Follow-up support

Below are brief explanations of each of these phases as implemented at many Air Force installations:

Phase I. Identification is exactly what the term implies. A member may be identified through civil or military arrest, apprehension, or investigation, or as incident to medical care. Similarly, a member may submit a urine sample which upon analysis proves positive for illegal or illicit drugs. Finally a member may voluntarily seek help through the Limited Privileged Communication Program (not discussed here, see AFR 30-19, paragraph 25). Regardless of the means of identification, once a member is identified in writing to the chief of social actions, that person is in the rehabilitation program, and this first phase is considered completed.

Phase II. Detoxification is the reduction of toxic properties or the changes which a foreign substance or chemical undergoes in the body to make it less poisonous or more readily eliminated. In practice, this is the "drying-out" stage and may or may not be accomplished while the rehabilitee is sedated. This phase is under the complete control of the attending physician. This phase is not always required and may be waived at the discretion of the attending physician.

Phase III. Medical evaluation and treatment are mandatory for all identified substance abusers. Medical treatment is provided as indicated. The purpose of this phase is to determine medically the appropriate treatment for each individual. Practically speaking, this phase is very short unless biological complications develop.

Phase IV. Behavioral reorientation is nonmedical and is designed to redirect the behavior of the individual so that he or she voluntarily conforms to Air Force standards and performance. Both individual and group counseling may be considered appropriate. It is my opinion that this phase contains the crux of the rehabilitative effort in that it deals with the psychological dependency upon the chemical substance or upon the effect produced by the chemical substance.

Phase V. Follow-up support is the process by which successful rehabilitees return to normal duty. The purpose of this phase is to monitor and facilitate the reentry of rehabilitees into normal military life and to help them to avoid a return to drug abuse.

I should now like to share with you my personal biases as they apply to rehabilitation. More specifically, I will expand upon Phase IV, Behavioral Reorientation.

AS I SEE IT

The therapeutic approach described below represents the culmination and current evolutionary stage of some three years of counseling. It dovetails with Phase IV outlined above and is offered here as a conceptual framework that has worked for me. If the reader can use it or adapt it in part or in total, he or she should feel free to do so. This theoretical foundation is original only in that it is a unique assimilation of well-known parts.

Basic Premise

The people in rehabilitation are rational adults who make decisions based upon available information. This premise presupposes that detoxification has already taken place and that behavior is no longer drug-affected. It also generates the following basic assumptions: 1. Rehabilitees are rational; 2. rehabilitees are adults; 3. rehabilitees are human.

If the therapist accepts these assumptions, the following must be accepted as *given:* 1. The client should be treated in a humane manner; 2. the client should be treated with dignity; 3. the client should be treated with respect.

Now let us examine these suppositions in depth.

I have said that the people in rehabilitation are rational adults who make decisions based upon available information. Right away, it becomes obvious that we are talking here about human beings who by virtue of their humanity possess human dignity and therefore deserve the counselor's respect. Permit me to inject some reality here. Respect is one thing, and complete abdication of the professional relationship is quite another. I recommend that the therapist respect the client, but I do not recommend that the therapist-client roles be abolished. Back to the basic premise. Let us examine the term, "rational adults." As with any group of people, the ability of rehabilitees to function rationally will vary. My contention is that all rehabilitees do, in fact, act according to the laws of reason, but to varying degrees. Father Joseph Martin, S.J., in his movie *Chalk-Talk* tells us that about 7 to 10 percent of all alcoholics need mental-health care, and this figure does not vary significantly from the national norm. Therefore, the challenge to the drug counselor is not to restore sanity but rather to increase the extent or degree to which the client uses reason. Thus the therapist is in fact dealing with rational people who do, as transactional analysis tells us, function at least sometimes as an Adult. If this is true, the rehabilitee does make decisions based upon available information.

Having laid the basic foundation, let us now consider a therapeutic model: Johari's Window (See Joseph Luft, *Group Processes,* Chapter 3).

	Known to Self	**Not Known to Self**
Known to Others	**Public Self**	**Blind Area**
Not Known to Others	**Private Self**	**Unknown Area**

The main theory behind Johari's Window is that personality can be divided into four broad categories depicted in the diagram. The Public Self is known to self and also known to others. The Private Self is that part of the personality which is known to self but not disclosed to others. This might include such things as a hatred of snakes or a basic belief that gynecologists are little more than sexual perverts who are licensed to practice medicine. The Blind Area, frequently referred to as the bad-breath area, contains such traits as body odor, overbearingness, and immaturity, which would be known by others but rarely known by the individual. The last area is the Unknown Area and it is exactly that — unknown.

The arrows in the diagram indicate possible sharing of information. If data are moved from the Private Self to the Public Self, the sharing is known as self-disclosure. If the cognitions are moved from the Blind Area to the Public Self, the sharing is known as feedback. In either case, more information becomes available to self, who in this situation is the client. Having accepted that the client makes decisions based upon available information, increasing the available information should improve the decisions. Thus the challenge to the counselor is to help the client take greater risks in the realm of self-disclosure and feedback. The National Training Laboratory tells us that the following criteria are useful for feedback and self-disclosure: (1) descriptive rather than evaluative; (2) specific rather than general; (3) considerate of the needs of both the giver and the receiver; (4) directed toward behavior over which the individual has control; (5) solicited rather than imposed; (6) always checked to insure communication. To this advice, I can add only that sincere caring has been the essential ingredient for me. With genuine concern, much can be accomplished and that which cannot might best be ignored.

Having established a point of departure and a *modus operandi,* we should consider a goal. Just what is the objective of this effort? The goal of rehabilitation as I see it is for the client to develop a nondrug-dependent lifestyle with which he or she can be reasonably happy. Two key factors deserve attention: "nondrug-dependent" and "reasonably happy."

Let us consider the concept of nondrug-dependence. It carries the connotation of some drug use, as I believe it should. If therapy relating to drugs (including alcohol) involves people, then it must also relate to reality. The American society of the 1970s is by no stretch of the imagination drug-free. We have caffeine in our coffee, zanthine in our tea, nicotine in our tobacco products, and heaven only knows what in our cold remedies and pain relievers. For most Americans, our lifestyle is neither drug-free nor dependent upon any particular chemical substance. Is this not what we should expect of our rehabilitation clients also? I contend that this at best is what we get and is what we should strive for. Drug-free is wishful thinking, not reality.

And what of reasonable happiness? It is a self-evident fact, which I will make no effort to prove here, that, if you remove the addiction from an individual and if the individual cannot find happiness without it, the person will return to the habit. In three years of counseling, I have never known this to be inaccurate, irrespective of the addiction (alcohol, narcotics, food, gambling, and work).

This then is rehabilitation (Phase IV) as I see it. The basic premise is that people in rehabilitation programs are rational, adult human beings who make decisions based upon available information. The superstructure is Johari's Window, and the goal is for the rehabilitee to develop a nondrug-dependent lifestyle with which he or she can be reasonably happy.

REHABILITATION GROUPS

As pointed out earlier, members of the military may be placed in rehabilitation involuntarily. The assets and liabilities of such forced entry will not be discussed here, but suffice it to say that forced entry does occur. This fact is indisputable. In the chart below, I have outlined what I have observed to be the basic and most consistent differences between forced-group members and voluntary group members with respect to initial participation in rehabilitation:

Voluntary Groups	*Forced Groups*
1. Attendance is directed by self.	1. Attendance is directed by an authority figure.
2. Participation is not known by authority figure (boss) and no label or stigma affixed.	2. Participation is known by an authority figure (boss), and individual is labeled with a negative stigma.
3. Some cost to client; therefore some personal investment.	3. No cost to client; therefore little or no personal investment.
4. Failure to successfully complete has little or no imposed effect.	4. Failure to successfully complete has imposed, negative effects.

A careful study of the chart yields the following working definitions:

Voluntary group

A gathering of persons who come together of their own free will for the purpose of mutual need fulfillment and have minimal outside forces influencing their collective behavior.

Forced group

A gathering of persons who are brought together by outside directive authority for the purpose of behavioral change and have significant outside forces influencing their collective behavior.

These working definitions are to some extent observable in the dynamics which they generate.

The literature tells us that groups develop and evolve in an observable and predictable manner. It has been my experience that the evolution of voluntary and forced groups does, in fact, develop in a phased manner; however, the phases differ significantly, depending upon the nature (forced or voluntary) of the group. Merle H. Ohlsen, in *Group Counseling,* relates the group development as observed by Thelen and Dickerman. My own experiences and observations agree with their findings..

VOLUNTARY GROUP DEVELOPMENT

Individually centered, competitive phase

Members try to establish themselves in the leadership hierarchy, or pecking order. They want a strong leader to take over and accept responsibility for them.

Frustration and conflict phase

When the leader fails to take over, members feel hostility toward him. He is perceived to be inadequate, inefficient, and the cause of failure and personal frustration. The members tend to blame others rather than accept personal responsibility for developing relationships in which they can achieve personal and group goals.

Group harmony phase

During this phase, cohesiveness develops, but it is accompanied by complacency, smugness, and "sweetness." Members are mutually supportive but tend to avoid or gloss over conflict and are not very productive. They curb impulse, especially negative reactions, and try to repress individual needs to satisfy group needs.

Group-centered, productive phase

Members still exhibit concern for others but not to the degree that they will ignore or gloss over conflict in order to achieve harmony. Members face conflict and learn to deal with it constructively. They accept responsibility for their behavior, participate in solving their group's problems, and develop productive working relationships. The group members also develop increasingly greater tolerance for others' values and behaviors.

These are the developmental phases of a voluntary group. I now contrast them with what I have observed to be the developmental phases of a forced group.

FORCED GROUP DEVELOPMENT

Denial phase

The members deny the existence of any problems of any kind with regard to themselves. They question the ability of the facilitator and blame the power structure for all inequities, including their current situation. Some group members may agree to "stay to help others," while clinging fiercely to their own lack of problems.

Group-harmony phase

This phase contains all the characteristics of the same phase in voluntary-group development, in addition to which it frequently evolves into a we-they situation, where the entire group tries to engage the leader in a win-lose or zero-sum game. If the facilitator is drawn into the game, all progress will cease.

Frustration and conflict phase

This phase contains many of the characteristics of the same phase in voluntary group development; however, it also contains the recognition by group members that they do indeed have individual needs and problems. This realization simultaneously marks the end of the denial process and the onset of positive, internal motivation toward personal change. It is essential that the leader be nonthreatening and nonjudgmental during this time, and one-to-one dialogue should be avoided. (By this I mean that the facilitator should turn questions back to the group and not allow the situation to degenerate into an attack on individuals.)

Individual-centered, competitive phase

Acceptance of the reality of the situation is now complete, and group members compete for the group's attention to their own problems. They seek a pecking order and a strong leader from within the group. The facilitator can easily assume a leadership role at this point but should avoid doing so in the interest of group growth. Trust the group; leaders will evolve, and the leadership role will rotate within the group, depending upon the particular short-term situation.

Group-centered, productive phase

This phase is essentially equivalent to the same phase in the voluntary group's development and has the same overall positive results.

It has been my observation that voluntary groups and forced groups commence with significantly dissimilar motivations and starting points, that both develop in a stepwise progression which is in no way correlated. I submit that the forced-group counselor needs to be aware of this basic and thoroughgoing distinction so that it may be positively encountered and productively channeled. I personally feel strongly about this and choose to share it with you, the readers, so that you may avoid the frustrations that I faced.

QUESTIONS FOR FURTHER CONSIDERATION

1. Are the present local, state, and federal laws relating to marijuana use realistic?
2. Should there be standardization throughout the United States for penalties related to drug use and drug sale?
3. Are present programs to correct drug abuse suffering from disagreement between those who advocate a purely physiological approach to the problem and those who advocate a purely psychological approach to the problem?
4. Should more emphasis be placed on early identification of those likely to abuse drugs?
5. Is a great deal of money now being wasted on ineffective drug education programs in the military?
6. What agency would be best suited to establish rehabilitative workshop sessions in the areas of alcoholism and drug abuse?

part 6
Health Care

We have left undone those things which we ought to have done, and done those things which we ought not to have done; and there is no health in us.

—*Book of Common Prayer*[1]

Where health is concerned, one human ideal has remained constant from ancient to modern times: a sound mind in a sound body. Despite the physical and psychological impediments of advancing industrial technology, which have polluted both the "inner" and the "outer" space of man, medical research has brought us tantalizingly close to this goal. Awesome improvements are reflected in diagnostic procedures, surgical techniques, and the broad advances in prophylactic and therapeutic medicine. Unfortunately, the delivery of health services has not kept pace with medical research, despite incredibly large expenditures. This is especially true in low-income communities, where a disproportionate number of military recruits reside.

THE NATION'S HEALTH

Health care is the nation's third largest industry, employing 4.4 million persons, including 320,000 physicians and 748,000 nurses. In 1973, the nation's health bill came to $94.1 billion, 40 percent of which was in tax dollars. That comes to $441.00 for every person in the United States.[2] Despite the fact that the United States spends almost $100 billion a year on health, uncalculated numbers of our citizens suffer and die yearly from infirmities that modern medicine could prevent, mitigate, or cure if medical resources were unrestrictedly available.

No contemporary event holds more citizen interest presently than this contest of health care versus neglect, where the ultimate stake may be life or death. The current state of health in this nation has reached a crisis level that is bringing increasing public demand for (1) easier access to health care; (2) increased medical personnel and facilities more equitably distributed; and (3) a national system of health insurance. Health, once considered a private and personal problem, has now become a community affair. In commenting on the crisis in America's health care, Senator Edward Kennedy observed, "Smaller countries spend smaller amounts on health care than the United States, and they give more health care to their people. In many of these nations, fewer children die than in America, fewer women die in childbirth, and men and women live longer lives on the average."[3]

Two of the indices most widely and frequently used to compare the health of Americans with that of other nationalities are the neonatal mortality (the number of babies who die in their first year of life), as well as female and male longevity. In a comparison of funds expended and results produced in health care, Americans do not fare well. Critics often point to statistics comparing infant deaths during the first year of life (America is 13th among the industrial countries), and life expectancy (America is 18th for men and 11th for women), as indicators of the overall quality of our medical care system.[4] As a specific example of quality and cost, the United States spends a slightly higher percentage of national income on health care than do Sweden and the United Kingdom.

> *. . . in one recent year the United States spent 6.8 percent, Sweden 5.6 percent, and the United Kingdom 4.9 percent. Yet the infant mortality rate is lower in England and Sweden, and the life expectancy is higher. An American woman can, statistically, expect to live two years less than a Swedish woman and a few months less than an English woman. The differences in longevity for men are even more striking: 67.1 years in the United States, as compared with 68.6 years in England and Wales, and 71.9 years in Sweden (which has the highest male and female longevity in the* world).[5]

Making a further international health comparison, the distinguished American heart surgeon Dr. Michael DeBakey, long familiar with Soviet medicine, says that while Russia lags behind the United States in the quality of medicine practiced, she does a better job of taking care of the average citizen. Asked in an interview to compare the accessibility of medical care for the average citizen in Russia and the United States, DeBakey observed: "Well, I think care is more accessible in Russia in the sense that the whole population has accessibility to it. It's accessible to everybody. In other words, they don't discriminate on the basis of finance in any way."[5] This is obviously not the case in America.

Nonwhites in particular are handicapped by poverty, discrimination, and social-psychological barriers that inhibit them from using the services that are available. The "Spanish" health problem has been intensified by the fact that large blocs of Spanish-speaking persons have immigrated to the United States, and epidemics and the language barrier have complicated their health problems. A Colorado study determined that neonatal deaths were three times as high among

Spanish-speaking groups as among Anglos, reflecting not only less adequate conditions at, and following, delivery, but also a lack of prenatal care, as well as the poor general nutrition and health care of the mother.[7]

A study of Puerto Ricans showed that Puerto Rico had a higher death rate from tuberculosis than any other country that gathers statistics. This susceptibility to tuberculosis is often exacerbated by poor, overcrowded housing conditions in the United States, as well as by deficient knowledge about health care and sanitation practices.[8]

Commenting on the health care of American Indians, one authority says that, although Indians of today are much better off than they were just a few years ago:

> *. . . they still have a long way to go to meet the standards of the more affluent Americans. Cultural conflict is still a problem, and the anomie and feelings of hopelessness created by this conflict are still evident. In the long run the only effective way to overcome the remaining health problems is to deal with base causes, starting with the problem of poverty.*[9]

Military recruits from poverty-stricken backgrounds often do not take advantage of free medical care. They must be taught to value and seek out good health care.

The comparisons that reveal a negative international picture for the United States in the delivery of health care, although disappointing, lack the impact of statistical data that show the disparity of health service among various segments of the population within the United States:

> *No finer medical treatment in the world may be found than what is now available at the Mayo Clinic in Minnesota, Johns Hopkins Hospital in Maryland, Methodist Hospital in Texas, Columbia-Presbyterian Medical Center in New York, or Stanford University Hospital in California. Yet only a few miles from these great humanitarian institutions, large segments of the nation's population receive hit-or-miss health care, with infrequent visits to doctors or public health nurses and virtually no dental treatment. The poor customarily enter a hospital only when they are very ill, often beyond help.*[10]

Illustrative of these differences in the health situation is a slum neighborhood in Boston where infant mortality exceeds the level of the Biblical plague inflicted upon ancient Egypt, in which one out of every ten newborn infants died. Describing this neighborhood, Selig Greenberg wrote:

> *The infant death rate of 111.1 per 1,000 live births, or one of every nine, in this section is five times greater than the national average, more than four times the average for the city of Boston as a whole, and more than 15 times the average for a nearby affluent suburban community which has only 7.2 infant deaths in every 1,000. In several other Boston ghetto areas infant mortality rates run from double to more than three times the city-wide average.*[11]

Although this is but a single example, it illustrates the paradox that, while health care is superb for one class or section of the United States, it is fatally inadequate for other portions of our citizenry. This legacy of neglect is one strong contributing factor for the symptoms of violence that are prevalent in our culture today. Rollo May cautioned that "such neglect leads to attitudes of personal insignificance, and the struggle for it, that underlies much violence."[12]

Family doctors have virtually abandoned Harlem, forcing local general hospitals that are already overcrowded and understaffed to undertake the treatment of all ills, however trivial. Other cities report the same problem. By contrast, the waiting room of the municipal hospital in San Diego is all but empty at times because many would-be patients are so disgruntled with red tape and staff shortages that they prefer to cross the border for treatment in Mexico. As a result of such inadequacies in health services, "Some estimates place the number of unfilled, decaying teeth in the United States at 1 billion. About 5,000 American communities do not have a single doctor, while over 100 *whole* counties lack the services of a physician."[13] Efforts to fill these gaps have led to a variety of practices — some of which may offer a remedy more hazardous than the condition.

A report prepared by five American physicians and published in the *New England Journal of Medicine* reports that a medical underground of thousands of unlicensed foreign-trained doctors, many of them unable to pass state examinations, is filling hospitals and clinics of the United States:

> *Eleven states knowingly allow the hiring of noncertified doctors for work in state mental hospitals and almost all states use* ECFMG *(Educational Council for Foreign Medical Graduates) certified* — but unlicensed doctors — *in a variety of roles as physicians with patient care responsibilities.*[14]

Some of the foregoing statistics and practices characterize the state of the nation's health and substantiate the assessment of experts that we have reached a crisis situation in health care: "In the ideographs of the Chinese language, two characters are used to write the single word 'crisis' — one is the character for 'danger' and the other is the character for 'opportunity.'"[15] This combination of the terms "danger" and "opportunity" would appear to be realistically descriptive of the situation that confronts those responsible for military health-care facilities. One of the first requisites for solution of this dilemma is a clear statement of the problems that have brought us to the present state of crisis in health care today.

MILITARY AND CIVILIAN HEALTH PROGRAMS

Owning more than four hundred hospitals with more than 180,000 beds, the federal government provides direct medical services to millions of persons each year. Patients in federal hospitals fall into four categories: (1) military personnel and their dependents; (2) veterans; (3) American Indians; and (4) federal employees with occupation-related injury or disease.

The first American military-medical department, the Army Medical Department, officially began in 1775. Since that time the Navy and Air Force have added their own health and medical programs. The military medical programs differ from civilian health programs in three areas:

First, complete health examinations, immunizations, and inoculations are routine procedure for military personnel and, under certain circumstances, for their dependents. In addition, preventive health measures are utilized as the chief means to combat health hazards (such as tropical diseases) that might occur in world-wide military operations and personnel transfers. Thus the emphasis on preventive health measures in the armed forces is unparalleled by any civilian program.

Second, through an elaborate screening process of examination and evaluation, most persons physically or mentally unfit for military duty are excluded. Those military patients whose disabilities make it certain that they will never return to duty yet who still require medical care are routinely transferred to the Veterans Administration. As a result, the active-duty beneficiaries of military health services are relatively young and healthy.

Third, the civilian patient can and, in the face of today's rising hospital costs, sometimes must convalesce in his own home. The hospitalized military patient generally remains in the hospital until he can be returned to duty. It must be cautioned that this is true only when the military patient is living away from any relatives or family (as in a combat zone) or has his living quarters on base, where, living with other military personnel on duty, no one will be able to take care of his needs full time. If the military patient (or dependent patient, for that matter) has any family living near the military installation, it is likely that the patient will want to be sent home to recuperate (provided his medical problems so allow).

The beneficiaries of a military health program, such as that of the Department of the Army, often include many people not serving directly in the military. Persons *entitled* to Army health service include all active-duty personnel, receiving their care as a matter of *right:*

> *Every* [*person*] *in the Army who is ill or wounded is entitled to thorough diagnostic study and treatment without charge, and may have the benefit of diagnostic procedures and advice to all consultants necessary for complete care without regard to expense, . . . remaining in the hospital until he has totally recovered, and receives full pay during this period.*[16]

Other groups of people are eligible for service and are served only if personnel and facilities are available after the active-duty personnel have been cared for. These "eligibles" include the dependents of active-duty personnel, certain retired personnel, and beneficiaries of other federal health services, such as the Commissioned Corps of the Public Health Service.

Whatever his classification, the patient's medical care in the continental United States is usually provided in one of three main categories of treatment facilities:

> *1.* Hospitals, *which provide relatively complete diagnostic and therapeutic services;*
> *2.* Infirmaries, *which provide beds and treatment for patients from a local command who have relatively minor illnesses or injuries; and*
> *3.* Dispensaries, *which provide outpatient services for military personnel only. These patients are treated or observed on duty, and also are excused from duty but are returned to duty during the same calendar day.*[17]

Generally infirmaries and dispensaries are integrated with the hospital system so that patients can be readily transferred to facilities with complete services. Overseas facilities are similar, except that patients requiring prolonged treatment are transferred to the United States as soon as their condition permits.

The primary concern of military hospitals is that competent health specialists care for the patients. All that needs to be treated by government physicians seems to be the body, not the total spirit of the afflicted man. One doctor is interchangeable with another and seldom do the patients see the same physician twice. The patients soon learn that each doctor operates differently. With such frequent changes in personnel, the patient cannot anticipate who will treat him on a particular visit. The professional relationship becomes unpredictable for the patient, outwardly losing its importance.

But inwardly the patient would still like to know and be treated more personally by the doctors. One social phenomenon recognized at both military and civilian clinics that seems to prove this idea involves the way the patients gossip about the doctors in the waiting areas. The patients, naturally wanting to know more about "their" physician (or possible physician), collect and trade stories about relatives and friends who have seen this doctor or that one. One tries to find out as much as he can about all the doctors. The exchanges among the waiting outpatients not only provide a kind of therapy for the nervous and anxious but also seem to make the doctors more familiar and personal, although in many cases they really are not.

Health professionals in federal employment, such as those in military hospitals, operate under different conditions from those in private practice. They have no professional office or clientele except those provided through their government jobs. They have little to say about whom they see as patients and what schedules they will work. Unlike civilian health professionals, they have offices in a hospital and spend most of their time attending not a number of familiar patients but endless crowds of faces, faces that change with each government transfer. While working in military hospitals, the health professional's concern for individualized patient attention is frequently given secondary priority.

Samuel Bloom explored some pertinent issues involved in military clinics when he talked about "the full context of the doctor-patient relationship."[18] Bloom diagrammed many of the factors affecting the relationship and the patient. His major hypothesis was that the physician and patient are not interacting in a vacuum. Doctor A has his major source of reference in the medical profession, while patient B has his major reference in his family. But behind these are the larger subcultural references of their individual backgrounds and the dominant cultural style of their society. A doctor is at a severe disadvantage in relating to his patients if he is unfamilar with their subcultural references (as in the case of Indian health care), just as the patient feels confused and upset when the organizational procedure of a clinic has different objectives and values from those to which he is accustomed.

The "instrumental interaction" (concerning the objective application of medical aid by the doctors) seems to be competently carried out in the government clinics. The "expressive transaction" (the human interaction level of the relationship), however, needs to be more deliberately developed in such clinics. A health professional may be an expert diagnostician, but, in dealing with patients, he needs also to be a human relator.

INTERPERSONAL RELATIONSHIPS

An important figure in the hospital organization is the nurse. One of the professions of nursing partly arose out of the practice of hiring servant girls to care for the wealthy sick and perform tasks that illness prevented them from accomplishing. Many nurses feel that there has been little change from those early attitudes among patients or the professionals they work with regarding their status as nurses and their capabilities. Unfortunately, the nursing profession has been identified with the history of the female, and a mother surrogate and healer label has made her heiress to many of the legacies of inferiority that have been attached to women generally.

Much of early-day nurse training was accomplished by apprenticeship, and this relationship emphasized what was essentially obedience training. A part of the current state of strain and unrest within the nursing profession derives from the practices many doctors — civilian and military — still have of commanding nurses and expecting their automatic and unquestioning response. Nurses are also growing increasingly critical of the tug of war between medicine and management for their time and allegiance in many hospitals. An indication of the dissatisfaction within nursing is reflected by the number abandoning it as a life work: in 1969 "close to one-third of those who maintained their registration were not employed in nursing."[19] Dissatisfaction is also registered by the growing participation of American nurses in their national association, which is working to bring about change.

Some of the negative factors that influence the relations between nurses and those they work with are lack of opportunities for nurses to exercise professional judgment. Many would prefer situations where they can function more or less on their own, responding directly to patients' needs, and deciding when intervention by a physician is appropriate. Too many hospitals, they feel, make it the sole prerogative of the physician to determine whether patients can have their hair washed, the extent of their recuperative exercises and the scope of the information they receive."[20] They feel that nurses too often have a position of life-and-death responsibility without parallel authority. More and more often the question is being posed whether all the areas currently under medical control are best situated in the physicians' domain. Nurses feel that too often they are hampered by the fact that the delivery of services in which they are expert are substantively controlled by those who, though they ought to set objectives, could not possibly be expert in all phases of delivery.[21]

On the national scene, in recent years, there have been a number of organized attempts to explore and improve relations between the physician and the nurse:

> *These conferences were useful and important; however, they remain gestures in the groping and exploring which characterize the relationship between the professions. Similarly hospital administrators, either within the orbit of the American Hospital Association or as members of the National League for Nursing, have gone through a decade of probing changes in the relationship with nursing without giving up much control of many management prerogatives.*[22]

Two very recent examples of the interpersonal misunderstandings and conflicts that plague the health-care system are the summer, 1974, nurses' strike in San Diego, which almost immobilized the health care of that area, and demonstrations of Vietnam veterans in Washington, D.C., one of whose basic demands was review and rectification of abuses of servicemen because of inadequacies in Veterans Administration hospitals. In both of these confrontations, lack of meaningful communication among the involved parties was apparent.

This lack of communication between the health service profession and those they serve has also reached a state of near deterioration for a large segment of the patient-doctor relationship. Few interpersonal relations are more vital in achieving positive results than those between doctor and patient, and yet there is growing dissatisfaction, especially among minority groups and the poor, about the health services they receive.

> *All morbidity studies made in the last 30 years have shown a greater volume of illness in the lower income groups. This is reflected by higher heart attack rates, longer average duration of illness, more days of disability and also higher death rates than the higher income groups. At the same time, the receipt of medical care moves in exactly the opposite direction; lesser quantities and quality of care are received by lower income groups.*[23]

A psychological factor that has bearing upon this patient-doctor relation has been assessed to derive partly from the respective images which patients and physicians hold regarding each other:

> *Ward patients recognized their low status and expected to be used as teaching subjects and, to some extent, as research material. The doctors reciprocated these expectations and*

> *offered little apology since these patients, though being charged for their hospital services, were not being billed for their medical care.*[24]

Another factor hindering a cooperative relationship between doctors and patients is the fact that minority and poor patients often view themselves as uneducated and incapable of understanding the explanation of professionals, whom they describe as "talking among themselves" and "using their Latin or their Greek." A great number of these patients feel "excluded" from their own illnesses, convinced that doctors deliberately and needlessly withhold information from them. Being embarrassed by their ignorance, patients often fail to put questions to their doctors and instead express overt behavioral hostility at being "pushed around" and ignored as persons.

The job of the military medical practitioner is further complicated by the belief of a number of his patients in "folk medicine concepts which are different from those of modern scientific medicine."[25] This is particularly true of minority and poor patients, who are often operating in distinctly different cultural mores. Meaningful relationships are still further complicated by the marked increase in specialization and the growing use of the medical-team approach, which makes the individualized pattern of the doctor-patient relationship an increasingly difficult one to maintain. Among the military, the doctor-patient relationship is further fractured by the necessity of minority and poor patients to use military health clinics, which means that patients are frequently cared for by different doctors and nurses on each visit, who often do not even know their names. One authority observes:

> *It appears that the degree to which the qualities ideally defined as essential to the therapeutic relationship, namely mutual trust, respect, and cooperation, will be present in a given professional-patient relationship varies inversely with the amount of social distance. Conversely the greater the social distance the less likely the participants will receive each other in terms of the ideal type roles of professional and patient, and more likely they will perceive each other in terms of their social class status in the larger society.*[26]

There is fairly general agreement that the low-status military patient, because of certain attitudes, has markedly less chance of obtaining optimal care than, say, middle-class-oriented officers. Physicians attribute this to several factors: the patient gives less attention to symptoms, he is less willing to sacrifice immediate for future gain, and he displays less individual responsibility. Physicians make their contribution to the often negative relationship between themselves and minority and low-ranking patients by being inflexible and failing to adjust to patient norms. Also the physician:

> *. . . sees the patient not only as a patient but as a member of a low-status group. It appears that this can make the physician less understanding of and less interested in the patient. Though the picture is obviously complex, one fact emerges. The low-status patient is less likely to have as meaningful doctor-patient relationship and as good over-all medical care as is his higher-status brother.*[27]

The role the hospital plays does not mitigate the strain of this physician-patient relationship.

The basic institution of health care is the hospital, which in many respects contains within its walls a microcosm of the world outside. Most people in the United States are born in a hospital, and great numbers of them will die there. Both patients and staff eat and sleep there, church services are held there, courts convene there, families gather there, and the direction of governments has been influenced by events that have transpired there.[28] Hospitals play an important role in building America's health and thus increasing the strength, productivity, and progress of our nation.

Although the patient is the primary consumer of hospital care and has the most at stake, he or she lacks knowledge to help bring about reform in this area of the health system that could be of benefit. Because civilian hospitals have been the chief offenders in the rising cost of health care, the attitude of many new military personnel toward them has become one of suspicious questioning. Until recently, the patient's perspective understandably has been personal and unconcerned with policy-making:

> *No amount of health education will produce a patient able to deal effectively with medical policy . . . but important effective expressions of consumer interest have come when the*

individual began to purchase care as a prospective patient and together with others entered into a continuous and collective relationship with medicine.[29]

The general hospital has been one of the most highly stratified and rigid of formal organizations. Like other larger bureaucratic institutions, it works toward achieving its goals through reliance upon such structural devices as a complex system of division of labor, an elaborate hierarchy of authority, formal channels of communication, and sets of policies, rules and regulations.[30]

In the hospital setting, certain structural mechanisms are required to effect a working relationship between these forms of authority. Conflict between authority segments inevitably colors many aspects of hospital life. Conflict between lay and professional lines of authority results from a different orientation and hierarchy of values, one emphasizing the maintenance of the operation of the organization, the other emphasizing the providing of services.

Civilian and private hospitals are not unique in offending those they serve. There has been an avalanche of criticism about some of the practices of Veterans Administration hospitals. Senator William Proxmire of Wisconsin said on July 7, 1974, that hundreds of thousands of dollars of medical equipment are sitting idle in Veterans Administration hospital facilities. He said the VA has "demonstrated a distressing lack of control over the proliferation of specialized medical facilities that are superfluous to the needs of our veterans. They are underutilized, over-equipped and some of the services can be properly performed less expensively."[31] The negative relationship of both veterans and taxpayers to such misuse of funds and facilities is at once apparent.

Some of the abuses in health care in the military are racial in nature, and the United States Navy has begun a program to promote racial understanding at the Portsmouth Naval Hospital and the Tidewater Regional Medical Center. This program is called UPWARD (Understanding Personal Worth and Racial Dignity). It involves a minimum of twenty hours in race relations training for everyone in the command. The hospital and center military personnel officer described the training as similar to "encounter-type groups in that there is a maximum of free personal exchange between members. Outsiders are not permitted to observe for fear of inhibiting participants from freedom of expression."[32] After having attended the session for executives, he "termed the experience as being great. Everyone gained a considerable knowledge regarding the feeling of minorities and majorities. 'It was very enlightening to me to discover the background causes of why minorities act as they do and to learn that putting your hands up in the air means different things to different groups.'" More programs of this kind are under consideration by other branches of the military and in the private sector.

Effective Communication in Nurse-Patient Relationships

Lt. Colonel Elender E. Jackson,
Chief Respiratory Therapist
Wilford Hall USAF Medical Center, Texas

> **I know you believe you understand what you think I said, but I am not sure you realize that what you heard is not what I meant!**
>
> —*Author unknown*

Wars have been waged, entire communities have been destroyed, lifetime friends have been lost, companies have gone broke, jobs have been lost, and in general people have been made unhappy, frustrated, and confined to jail or to mental institutions as a result of ineffective communications. An excellent example is that, despite the billions of dollars the United States has spent on foreign-aid programs, the United States has captured neither the affection or esteem of the rest of the world. Many of our problems arise as a result of ineffective communications on our part.

DEFINITIONS

To paraphrase *Webster's Dictionary,* communication may be defined as a transmitting, a giving, or giving and receiving, of information, signals, or messages by talk, gestures, writing, and so on. *The American College Dictionary* defines communication as the imparting or interchange of thoughts, opinions, or information by speech, writing, or signs.

Considering the definitions of communications, we can readily see that communicating is a many-faceted art. Dictionary definitions stress that communication is many things, including the spoken word. To be or become effective communicators, nurses must first of all know what communication is and must understand other persons — their culture, gestures, body posture, language — that is, the words a person uses and in what context, the presence or absence of vocal inflections, and facial expressions. Another important task is learning to listen, to take mental shorthand of what a patient says, the signs and signals he or she gives, and to try to decipher the total meaning of the communication effort.

The nurse's greatest challenge is people, people of all ethnic backgrounds, cultures, and from all walks of life. It is necessary, then, that we realize that, because a patient is admitted to the hospital or is seen in an outpatient setting, the patient is no less an individual. He does not stop acting like a person, nor does he give up his personal idiosyncrasies or status when he becomes a patient. This is a fact that we so often overlook in dealing with the sick.

Hospital managers, doctors, and nurses should be more concerned about the patient as an individual. The nurse, who has more direct contact with the patient than any other member of the establishment, can do much in the way of recording and reporting on the patient's medical and mental condition. An astute, conscientious nurse, well versed in the art of communicating, can often detect personal problems that the patient may have that might have a direct bearing on his or her desire to get well and leave the hospital, or to malinger and stay within the haven for protection from his problems or from fear of the unknown.

KINDS OF COMMUNICATION

Verbal Communication

Verbal communications consist of words — words uttered by individuals in an effort to transmit information, to solve problems, or merely to engage in social intercourse. Words play an enormous part in our lives and are therefore deserving of the closest study. Words have the power to mold men's thinking, to canalize their feelings, to direct their willing and acting.

People of similar cultures have agreed that certain sounds, grunts, and gibberings we make with our tongues, teeth, throats, lungs, and lips systematically stand for specified happenings in our nervous systems. We utter these sounds in hope that the receiver of them is in common agreement with what the sounds stand for. Words are what make us human; their value is transcendent. At the same time, words are full of traps, distorting evaluation, leading to pain and misery beyond all reason. There is no escape from this problem. Words are helpful, harmful, neutral — in all shades and at all levels. The attitude to take, I believe, is thankfulness for the power and utility of words and determination not to misuse the gift. Words are like a sharp ax — invaluable for the pioneer as long as he does not let it slip. Today, however, sharp axes are flying in all directions.

Words and their meanings, in conjunction with how the nurse or patient perceives the words used, will determine how successful or unsuccessful we will be in verbally communicating with each other. Effective verbal communication in nurse-patient relationships is dependent in part upon the extent of words both can understand and the words each can use effectively, that is, words that both can understand and act upon. Here I allude to the specialized medical vocabulary that the nurse must know in order to communicate with her professional colleagues. But when she is communicating with patients, she must be educated enough to be able to talk to or converse with them on their level.

Nonverbal communication

Nonverbal communication is severely limited and should never be considered a substitute for words, but its skills are an adjunct to the effective use of language, and should be cultivated and improved to increase our overall person-to-person effectiveness. Nonverbal communication is commonly called "body language." Within the last few years it has been labeled the "science of kinesthesia." This science includes nonreflexive or reflexive movement of a part or all of the body used by a person to communicate an emotional message to the outside world. There are several kinds of body language that people use in conjunction with the spoken word. Some of the most common ones are discussed in the following paragraphs.

1. *Gestures and clusters of gestures.* It has been estimated that there are 100,000 distinct gestures that have meaning to people around the world. These gestures are produced by facial expressions, postures, movements of the arms, hands, fingers, feet, legs, knees, and so on, and by combinations of these. Gestures are essential in face-to-face communication to help us get our points across. They can be either positive or negative, indicating approval or disapproval.

2. *Manner of speaking.* The tone of our voices, the placing of oral emphasis, is closely related to the gestures we employ. The content (words) and the way we deliver — the quality, the volume, the pitch, and the duration of our speech — all have a strong impact on our listeners, favorable or unfavorable.

3. *Zones of territory.* Edward T. Hall has coined the word "proxemics" to describe his theories and observations about zones of spatial territory and how we use them. Zones represent the areas we move in, areas that increase as intimacy decreases. Under normal conditions, a distance of six to eighteen inches is considered distant, depending upon the persons involved. In informal gatherings, this distance would be considered too close for the average American male, whereas for American females this distance is acceptable. When our zones of territory are invaded or we in turn invade the territory of others, we have ways to let the individual know that we did not mean to invade his privacy. For example, if we bump against someone in a crowded bus, train, or elevator, we smile to say, "I'm sorry, I didn't intend to intrude, nor to be aggressive, I did not intend to hurt you." If we wish our territory not to be invaded, we use body positions to indicate that we wish to be left alone, or we can look at a person in such a way as to convey the same message.

4. *The eyes and eye contact.* The look or the stare can give a person a human or nonhuman status. In our society, we do not stare at human beings; we acknowledge their presence by using a deliberate and polite inattention. We look just long enough to make it clear that we see them,

and then we immediately look away. We are saying in body language, "I know you are there," and a moment later we add, "but I would not dream of intruding on your privacy."

5. *Touching.* The sense of touch also conveys messages. Sometimes it can and does elicit a negative reaction from the person being touched and from the one doing the touching. In nurse-patient relationships, it is imperative that the nurse touch the patient or vice versa at some time during the patient's hospital stay. The nurses should be aware that she, as well as the patient, can send out a negative message and that negative feelings between the two interfere with their relationship.

6. *Listening.* The missing link in effective communication is listening. Listening is one of the most valuable tools the nurse possesses. It involves hearing, recording, being empathic, and assimilating what the patient says. Observation of the patient's reactions to what the nurse says is also an important part of listening. The nurse can often tell whether the patient is really listening and whether the patient understands her message. The nurse must always be aware of the fact that the patient observes the nurse and can also tell whether the nurse is listening and understanding what he or she is saying.

The importance of nonverbal communication is easily demonstrated by the newborn infant. He or she is not able to read or to understand language. The infant is influenced by the nonverbal communication elements in the environment. The way he is held and fed, the way he is rocked and placed in his bed; the nature of his bed or crib, the colors, the lighting, the sounds that surround him — all these have an immediate and probably a lasting effect on the child. And never throughout life will he or she be able to avoid the impact of such nonlinguistic communication devices as actions, symbols, signs, gestures, mannerisms, sounds, and signals.

LEARNING THE ART OF COMMUNICATION

Observation, recording, reporting, understanding, empathy, and intellectual inquiry into cultures other than our own are the stepping stones to learning the art of communication. We should endeavor to:

1. Understand ourselves and the receiver of our messages. We

should be especially careful of the words we use, the signs and signals we ourselves transmit.

2. Speak clearly and briefly and at a pace that is not so slow as to lull our listener to sleep nor so rapid that the receiver cannot grasp what we are saying.

3. Use words that both we and the receiver understand.

4. If we are the listeners, ask for clarification if the sender's message is not clear. If we are the senders, search for gestures, signs, or signals from the receiver to let us know whether we are being understood.

5. Listen to try to hear what the sender is saying and try to empathize with the sender to understand better his or her message.

6. Understand the meaning of gestures and/or clusters of gestures. Use gestures to help get the message understood.

7. Be alert so that we can detect congruence or incongruence between what the patient says and what he does or how he behaves.

8. Learn how to read the signs and signals the transmitter uses.

9. Keep an open mind. Listen and try to understand what is said, learn how to report what we have seen, heard, or felt accurately without inferences and without passing moral judgment.

10. Increase our knowledge of other cultures and the ways in which they employ verbal and nonverbal communications.

Learning the art of communications is a challenge for anyone. However, it is especially important that nurses put forth every effort humanly possible to provide patients with the best nursing care possible. Our clientele must be understood from a physiological (medical), psychological (mental), and *cultural* point of view.

NURSE-PATIENT COMMUNICATION

Interviewing is a process that is carried out every day in almost every situation in which there is an exchange of information between two people. This exchange may be between a nurse and a patient, a doctor and a patient, a social worker and a client seeking financial aid or some other social service, or between many other individuals who are engaged in getting information for people, advising people, or helping people to live better and happier lives.

The nurse in a hospital situation interviews patients every day. A mere, "Good morning, Mrs. S. How are you?" or "How did you sleep last night?" are open questions that might lead to a flood of valuable information for the medical and nursing staff if the nurse notes what the patient says, how she says it, her behavior, her mannerisms, and her emotional state.

When the nurse admits a patient to the hospital, she conducts an interview. This initial interview may result in some valuable information that will be helpful in evaluating the patient and planning for subsequent medical or surgical care. This opportunity is not used to its greatest advantage. Most nurses, especially those trained in a traditional school of nursing, adhere to the hospital printed form, ask direct questions to make sure all the blank spaces are filled, and usually fail to allow the patient time for self-expression.

If the patient asks a question, the nurse usually gives long, detailed answers, which in most instances serve only to allay the nurse's fears or her feelings of inadequacy, thus leaving the patient floundering. The usual place for admitting a patient to the hospital and conducting the interview is at the nurse's station, where there is a continuous stream of traffic, or in the patient's room with an audience of two or three other people. These situations are not conducive to a good interview; they block the patient's desire to ask questions or to talk freely.

After the printed form has been filled out and the patient's vital signs taken, the nurse completes the written nurse's record. A typical example of an admission record is as follows:

> *Mrs. S, age 35, admitted to Ward 10 walking. Patient is to have a simple mastectomy in the A.M. B/p 150/100, T 98, P 120, R 28. The patient does not seem to be in acute distress, admission bath given. Patient to bed. Dr. M notified of patient's admission.*

What does this record tell us? Merely that we have Mrs. S on the ward, what her vital signs are, that she had a bath, that she was put to bed, and that the doctor was notified. If the patient showed signs of anxiety, if she seemed to be preoccupied, or if she expressed concern for her children or her husband, we would not know it by reading this record. Many patients, after having been ushered down the admission assembly line, having been denied the right of self-expression and kept away from family and friends except for a certain few hours a day, generally feel that they have lost their identity and their individuality and have become just another hospital case in bed three.

To conduct a successful interview, nurses in the hospital should familiarize themselves with the purpose of interviewing, methods of conducting an interview, the things to look for during an interview, and the essential physical setting.

Our first consideration should be to put the patient at ease and make him as comfortable as possible. This can be accomplished by introducing ourselves and any other ward personnel who are present at the time and by seeking an empty room or an office in which to conduct the interview. Second, we should explain to the patient that we will need to ask a few questions concerning his or her previous health status and that of his or her family and that this information will enable us to take care of present and future health needs. Since our primary purpose is to elicit information about the patient's health, it is necessary to ask questions, but we should keep in mind that it is important to let the patient do most of the talking. The nurse may need to direct the interview to a certain extent, but she should not dominate the interview, thus preventing the patient from expressing himself. Self-expression on the part of the patient may uncover hidden problems and fears. To discover these problems early in the patient's hospital stay is vitally important in helping him or her make a speedy and uncomplicated recovery. In a hospital situation, where there are doctors present to take care of the patient's needs, I think it is wise to let the patient talk without too much direction. I feel that the nurse should be a good listener and record any pertinent information the patient reveals.

Letting the patient talk has a special significance if he or she is about to undergo surgery and anesthesia. If the patient shows signs of emotional instability, it is well for the surgeon to know about the situation before surgery is performed. Perhaps the patient is in need of mental-health guidance and reeducation concerning his or her illness. In the case of the thirty-five-year-old woman who was to have a simple mastectomy, mentioned above, there could be many hidden fears about her future happiness, for after all this would be a very traumatic experience for her. To proceed with the surgical procedure without knowing her mental state could cause her to develop a full-blown psychosis, resulting in a prolonged hospital stay, loss of family and social status, and undue financial strain on the family, not to mention the mental anguish and deprivation that would be inflicted upon the family.

The nurse must be aware of what to look for in interviewing and what things are important to bring to the doctor's attention. Some of the things she should notice are: How does the patient associate ideas? Is there some previous experience connected with the ideas expressed? Is there a shift in the conversation? The nurse may notice a shift, but in the mind of the patient, it is a continuation of the same idea. Are there concealed meanings in what the patient says and how he or she says it? Are there inconsistencies and gaps in the conversation? Are the meanings clear? If the patient contradicts himself, this may indicate ambivalence, internal pressures, guilt, or confusion. All these things the nurse should note, record, and call to the doctor's attention.

Much has been said about what things the nurse should observe in the patient, but it is just as important that the nurse know herself. She must admit that she has prejudices, fears, anxieties, and ambivalent feelings and face up to them. If the nurse is to take her place in giving total health care and supervision to patients and their families, she will find it necessary to become an accomplished interviewer.

NEGATIVE COMMUNICATIONS: TWO CASE STUDIES

The *invisible patient* is one who is treated as if he or she did not exist or one who has been relegated to a nonhuman status. Medical technology has provided ways to keep people alive by the use of various drugs, cardiac pacemakers to keep the heart beating, machines to dialyze the entire blood stream to remove impurities that the body can neither excrete nor metabolize, and mechanical respirators to breathe for the patient if he cannot breathe for himself. The list could go on and on, but the ones listed above serve to illustrate my point.

Consider the patient who cannot adequately breathe. Various medical and surgical conditions can and do affect the respiratory system to the point where the patient cannot add adequate oxygen to his blood stream or remove carbon dioxide from it. Such a patient would be intubated (a procedure by which a tube is passed through his mouth or nose into his trachea to provide an airway and to remove secretions from the tracheobronchial tree) and put on a mechanical respirator which would provide the ventilatory support necessary. To ensure cooperation between the patient and the respirator, certain drugs are administered that will paralyze the muscles of the patient's body, including the respiratory muscles. These drugs will *not* obtund the patient's sensorium. Here we have a patient who is unable to move a muscle, and yet he can hear and understand parts of all that is said in his presence.

Mr. A is the patient described above. In addition to the respirator, the endotracheal tube, and the paralyzing drug, he is also connected to a continuous cardiac monitor for easy observation of his heart rate and rhythm, a central venous pressure line to monitor venous pressure, and a multitude of intravenous and intra-arterial lines. A typical conversation at Mr. A's bedside might be as follows:

NURSE: *(While suctioning Mr. A's airway calls to John). It's time to take an arterial sample on Mr. A.*
JOHN: *I'll get it right away. (The blood sample is drawn and run through the blood gas machine, and John returns to the bedside.)*
NURSE: *Gee! The PCO_2 is very high, the pH is much too low and the O_2 is going down. I think we should call Dr. R.*
JOHN: *Dr. R said to call R.T. to readjust the machine if this should happen.*
NURSE: *O.K., then call Dr. R and ask if we can increase his respiratory rate and tidal volume and if we can give an ampule of bicarb I.V. Then call R.T. and ask for a* PEEP *setup and to hurry because we need the equipment* STAT.
(The respiratory therapy technician arrives.)
R.T.: *I don't think we should put the* PEEP *on just yet. Let's increase the FIO_2, take another blood gas in 15 minutes and see what it shows.*
NURSE: *No, I think we should do something now because Mr. A is having V. Tach. and a few* PVCS.
(Dr. R arrives.)
DR. R: *How much fluid has Mr. A had within the last hour? Is that all the urine he has put out in the last hour? What are his vital signs? Get me some bicarb right away. When did he start having atrial fib? Call X-ray and get a chest X-Ray.*
NURSE: *Dr. R, if you would look at this lead two, you can tell that he is having V. Tach. and* PVCS *and not atrial tach.*

(The X-ray technician arrives.)

(Mr. A is lifted and turned this way and that to get the X-ray film under his back.)

X-RAY TECHNICIAN: *Everybody out! I'm ready to shoot.*

(The X-ray is taken, and Mr. A is again turned this way and that to remove the film. Everyone returns to the room and the treatments continue.)

Mr. A can hear everything that is said but can do or say nothing. By this time, he is in serious doubt about what is really wrong with him and whether he will survive it at all. Mr. A has been ignored; he is no longer a living human being. No one had the presence of mind to alert him about what they were going to do to him or for him. His condition should have been discussed elsewhere out of the range of his hearing.

The *distressed patient* is one whose recovery is inhibited by family, financial, or other personal worries.

Mr. M is forty years old. He has had numerous admissions to the hospital for treatment of a heart condition. He is married and the father of four children, all under twelve years of age. His income is a small pension allotted to him because of his heart condition. After a surgical procedure to replace his defective heart valve, Mr. M was able to perform light work on a part-time basis to supplement his income. He and his family also lived in a mobile home so situated that he and his wife could raise vegetables and a few pigs and chickens to help meet his family's needs.

Mr. M has been followed as an outpatient since his valve placement. When Mr. M told the doctors that he seemed to be having problems with his heart, he was admitted to the hospital for observation. After about a month of hospitalization, the doctors concluded that the valve was malfunctioning and would have to be replaced. My first encounter with Mr. M was five days after the surgical procedure had been performed.

Enroute to the ward to see Mr. M, I noticed a lady whom I had seen in the telephone booth several times. I had noticed that she seemed to be upset by the way she spoke into the receiver, her tone of voice, and her facial expression. At that time I did not know that there was any relationship between her and Mr. M.

I walked into Mr. M's room, introduced myself, and told him why I had come to see him. He was sitting up in bed taking a breathing treatment on a mechanical respirator. He was perspiring profusely, his pulse rate was quite rapid, and his face was flushed. I interrupted his breathing treatment to ask how he felt and how long he had been perspiring so. He answered that he felt like hell, was hot and then cold all over, and didn't understand what was going on inside him. He seemed quite uncomfortable and in distress, his respirations were rapid and shallow. I asked him if he could slow down his breathing and take deeper breaths to try to get more air into his lungs.

At this point, the lady I had seen in the telephone booth came into the room and sat down, frustration and helplessness written all over her face. I introduced myself to her, and she said that she was Mrs. M. Mr. M asked her what she had found out. Mrs. M did not answer. She shrugged her shoulders, and told him to finish his treatment. Mr. M resumed his breathing treatment until the therapy technician returned. He said to Mr. M, "That's enough now; you have been on for fifteen minutes." I told the technician that I was sorry that I had interrupted his treatment, that he had been off the machine for at least five minutes, and that if he could continue the patient's treatment for at least five minutes I would stay with him and take care of the equipment when he finished. The therapist agreed and left the room.

The ward nurse and ward technician came into the room. The nurse had two syringes and needles full of medication, one to give Mr. M in the muscle and one to add to his intravenous drip. The technician had sterile gloves, instruments, and gauze to change Mr. M's dressing. They flanked Mr. M, one on each side of the bed, to perform the tasks they had intended. The ward technician said to the nurse, "I think this thing is infected. Maybe that is why his temperature is so high." The nurse replied that the doctor did say something about an infection but did not know exactly where it was. It could be in the chest, the valve, or at the drainage site. They finished their chores and left the room.

After everyone had left the room, both the patient and his wife started talking to me. They talked about the many problems they had. Mr. M was worried because he didn't know what was causing his surgical problems. He had no idea when he would be getting out of the hospital, under what conditions, or how he would be able to support his family after he was released. Mrs. M assured him that they would make it somehow.

I told Mr. M that I would relay the information he had given me to the charge nurse and that she would get his doctor to come in to talk to him and that he should be sure to tell his doctor the things he had told me. When he finished his breathing treatment, I cleaned the equipment and started to leave the room. Mrs. M asked if she could come with me. I said, "Sure, I'm just going to speak to the charge nurse."

As we walked out of the room, Mrs. M said she wanted to cry and that she could not tell Mr. M what she had found out because she would start crying in front of him and she did not want to upset him further. Indeed, she did begin to cry, and I found an empty doctor's office where she could cry in private. I wanted to go and find the charge nurse, but she asked me not to leave her at that moment. Then she told me that she had some kind of seizure disorder and that she was on medication, so I went to get the water for her. In the meantime, I saw the charge nurse, and she came with me to tend to Mrs. M. Then she went out to find Mr. M's doctor.

When Mrs. M calmed down a bit, she also told me that the creditors were coming to take their mobile home the next day if she did not pay the three overdue payments, that their food supply was getting low, and that the utilities were about to be cut off because she could not pay the bills. She was worried, too, about her husband's condition and his recovery.

Situations of this nature arise more often than we wish to admit. We nurses are so busy trying to get medication to patients and giving necessary treatments that we really do not have the time to stop and listen to our patients. Lack of communication between the nurse and patient and/or his family often leaves hidden fears and problems the patient may have in addition to his medical or surgical illness.

A time to listen and a time to talk are applicable here. A few minutes of the nurse's time would have uncovered the home and family problems confronting Mr. M. By careful listening, recording, reporting, and empathizing with Mr. M and his wife, the nurse could have very early in the patient's course of hospitalization alleviated the additional stress and strain placed on this patient. His recovery might have been less stormy and less eventful.

CONCLUSION

As professional nurses, we must remember that sick people are essentially dependent people looking for security and that they find that security primarily in their feelings for, and attitudes toward, us as nurses. Consequently, just being the kinds of nurses who understand those feelings and attitudes is tremendously helpful to many sick and troubled and uncertain people. More than that, we give these people an opportunity to talk. We do this best when we listen and say as little as possible to permit the patient all the freedom he or she may need to share the various troubles with us, for it is in this talking and sharing that a large part of therapeutic usefulness lies. The patient is relieved of anxiety by the mere fact that for the first time he or she can talk freely about some things. We should not underrate listening as therapy. It is very important. It may seem relatively casual to us, but it is the very crux of therapy. It is an important function.

We go beyond that in dealing with some individuals and their internal problems. We have already learned that, as patients discuss these things, we find that they have strange attitudes, misinformed attitudes, that they are frightened, that they are completely at sea about what their illnesses mean. Part of our job, then, is the explanation of what these things really mean and how they will affect their lives. This can be very surprising, very new, to the patient, but it can also be tremendously valuable therapeutically, for in so doing we take all the anxiety and the sting away from the somatic or organic interpretation and put it where it rightfully belongs, in the area of the patient's personality. That very frequently brings immediate relief.

In addition to listening, we must also remember that the most effective communication is personal, and is two-way. We must understand that the spoken language and body language are dependent upon each other for understanding communication. We should not, nor should anyone else, depend totally upon either one as the total communicative effort. The spoken language is often incongruent with the patient's behavior or body language, and the reverse is also true. Congruence between the two is the key to understanding the messages being sent or received. The truly professional nurse will be cognizant of cultural differences and how other cultures handle both the spoken language and body language. She will conduct herself accordingly and plan her nursing care to fit the individual patient.

QUESTIONS FOR FURTHER CONSIDERATION

1. Is the American Medical Association serving the interests of doctors to the detriment of the patients' interests?
2. Who should decide the limit of the sphere of authority of doctors, nurses, and hospitals?
3. In whom should decisions on health care priorities be vested to best serve the interests of the majority?
4. Can the military feasibly be used to train a new corps of medical auxiliaries to relieve the overburdened health-care profession?
5. What methods can be devised to promote understanding and cooperation between alienated segments of the health-care system?
6. Can the schools play a larger role in the education and delivery of health care to our children?
7. Should human relations training be mandatory for all health personnel?
8. What are some of the innovative health care programs being utilized in the military?
9. Should all branches of the military make greater use of civilian health care facilities?

part 7 Civil Service

Democracy is still upon its trail. The civic genius of our people is its only bulwark, and neither laws nor monuments, neither battleships nor public libraries, nor great newspapers, nor booming stocks, neither mechanical inventions nor political adroitness, nor churches nor universities nor civil service examinations can save us from degeneration if the inner mystery be lost.

— *William James*[1]

When his work force includes both military and civilian employees, the military commander quickly learns that it takes considerable skill in human relations to accomplish equal-employment-opportunity goals. While he may be able to order his military personnel to carry out specific orders, he must follow United States Civil Service regulations and procedures when dealing with civil servants.

DEVELOPMENT OF CIVIL SERVICE

The United States government now numbers 2,000 agencies and employs more than three million people. Less than two hundred years ago, at its founding, it consisted of nine executive units employing 1,000 federal workers.[2] In the early days of our nation, President George Washington found little direction in the Constitution concerning the appointment of individuals to administrative office. Nevertheless, his policies in this area set the pattern for the years between 1789 and 1829, and his high ideals have remained a beacon to this day.

President Washington stated that the only consideration that would influence his appointments to public office was to be "fitness of character." One authority has commented that at the time "'fitness of character' could best be measured by family background, educational attainment, honor and esteem, and, of course, loyalty to the new government."[3] Although political parties came into being only a few years after Washington assumed office, it does not appear that party affiliation automatically disqualified one for a position in the federal government, at least not until 1829.

Quite early, the federal service was divided into two basic groups: an elite administrative group and a second group whose jobs were carried on in offices and in the field, that is, clerks, postal workers, surveyors, and customs employees. Most of those in the second category would now be covered by the federal civil-service system. The elite administrative group was naturally drawn from the socially and intellectually prominent people of the new nation. Members of the second group came largely from the middle and upper-middle classes, and their education and professional competence, while sufficient to perform the appointed tasks, were not comparable to those of the elite.

The concept of appointment to federal positions was greatly altered by the election of 1828, from which Andrew Jackson emerged triumphant. Although his election is said to have signaled the beginning of the spoils system, Jackson's characterization of federal service and federal servants in itself is difficult to quarrel with:

> *The duties of all public offices are, or at least admit of being made, so plain and simple that men of intelligence may readily qualify themselves for their performance; and I cannot but believe that more is lost by the long continuance of men in office than is generally to be gained by their experience.*[4]

There is no doubt that Jackson and those who followed him in the presidency severely weakened the power of the social aristocracy that had controlled federal appointment for the first forty years of the nation's history. Nevertheless, it is difficult to assert that the so-called Jacksonian revolution actually meant any great change for the better in the process of filling federal jobs. If anything, the spoils system made it more unlikely that those responsible for inefficiency and dishonesty in government service could be held to account for their misdeeds. "We had," as Frederick C. Mosher stated, "effectively though not completely transferred governmental power from one group (the gentry) to another (the politicians); in the process, we suffered a considerable degradation of public office and widespread corruption."[5] The situation that prevailed is characterized by the words of Thomas Hobbes, "For the nature of power is in this point, like to fame, increased as it proceeds; or like the motion of heavy bodies, which the further they go, make still the more haste."[6]

One of the basic purposes of early civil-service movements was to reform the representative system by lessening the use of public offices to reward support for party and factional leaders. These movements sought some test of merit or qualification for persons given civil-service appointments. A victory for reform was won in 1871, when legislation adopted as a rider to an appropriation act authorized the president to prescribe regulations for the admission of persons into the civil service. These regulations were to be structured to promote the efficiency of the country and additionally to ascertain the fitness of each candidate with respect to age, health, character, knowledge, and ability. The president was empowered to employ suitable persons to conduct such inquiries, to prescribe their duties, and to establish regulations for the conduct of persons who were to receive appointments. For this initiation of civil service, $25,000 was appropriated and President Ulysses S. Grant appointed an advisory board, later to be called the Civil Service Commission.

Although President Grant may have had the best of intentions in his attempt to reform the civil service, he cannot be said to have made a major step toward ending the corruption. It took the assassination of President James Garfield by a disappointed officeseeker in 1881 to bring to the attention of the public at large the serious need to reform the rules for government employment.[7] The Civil Service Act of 1883, commonly called the Pendleton Act, was the first great piece of legislation in American history designed to protect the federal civil service from inefficiency and corruption.

The act was influenced greatly by the Northcote-Trevelyan report in Great Britain, which reformed the civil service of that nation. However, two major provisions of the act were drawn up to avoid the class discrimination inherent in the civil-service system of Great Britain. First, though it was stipulated that entrance into the American civil service was to be dependent on a qualifying examination, it was also required that such examinations be "practical" in character. Second, in keeping with the democratic origins of this nation, it was set down that entrance into the system could occur at levels above the lowest grade; that is, lateral entry was not prohibited. The act further provided that all positions to be filled should be "analyzed, described, and related to the knowledges and skills required for their performance."

The Pendleton Act came at a time in American history when reform was much in the air and the people were tired of the rampant dishonesty abounding in the federal government. For the next thirty years, government at all levels in the United States was subject to the same spirit of reform that characterized the passage of this act. Perhaps its most important provision was the establishment of the principle of political neutrality for the civil servant, which promised a stable and continuous administrative service free of partisan pressures, obligations, and removal. The act was perhaps both a culmination and a beginning of a reform movement which drew a "clear distinction between right and wrong, between the 'good guys' and the 'bad guys.'"[8]

The establishment of civil-service commissions in the act tended to exert a twofold influence: they associated public personnel and administration with morality, and they divorced personnel administration from general management. The act established in the minds of Americans the idea that politics and policy should be separated from administration, an idea that has persisted to the present day.

After World War I, the emphasis in the federal civil service was placed on "efficiency" — to place the right man in the right job with the least amount of money possible expended to do so. There was a serious attempt to make government "more businesslike." It was thought that the forms, structures, and procedures useful in business could be equally beneficial in government. This is not surprising when one remembers that the 1920s saw the apparent triumph of the management theories of Frederick W. Taylor.

The success of aptitude testing programs in the Army in World War I convinced many in government that the examinations for civil-service appointment could be refined and made more precise. Promotions were given more and more on the basis of "efficiency ratings." "Position classification" became the watchword in civil-service management. As Mosher remarked, "Personnel administration was effectively depersonalized."[9] By 1930, the United States Civil Service Commission could boast that civil-service protection was enjoyed by four-fifths of the nonmilitary employees in federal service, as compared with only about 10 percent in 1884.

Perhaps the single most important event bearing on the direction of the federal civil service after World War II was the creation of the Hoover Commission. This body set to work to resolve once and for all the supposed conflict between the admittedly partisan operation of policy setting and the legally ordered nonpartisan duty of administration. The commission did not deny that, in response to the will of the majority of the people, each new federal administration had a right to place its adherents in policy-making positions, but it did strongly endorse the principle that there must be many trained, skilled, and nonpartisan employees in the federal service to provide continuity in the administration of the government's activities. Unlike its predecessors, the commission undertook to spell out the distinction between executive offices that would be politically filled and those that should be filled on a career basis within the civil-service system. The precise form and function of civil service is still today the subject of much debate and various interpretations.

FORM AND FUNCTION

One view contends that the civil service of a political body includes people employed to administer the civil programs of a legal unit of government, except for those elected or appointed to political-executive offices. A more popular but restricted view restricts civil service to those persons who have been appointed to civil-government positions after some nonpolitical test for fitness. Whichever definition is accepted, in modern times the importance of civil servants had increased markedly. In an address commemorating the ninety-first anniversary of the federal civil service, then Vice-President Gerald R. Ford said, "A strong career service is one of the greatest strengths of our democratic process, and one of the best guarantees of sound, effective and efficient government — even more so in 1974 than in 1883."[10]

As the bureaucracy of government increases with the growing complexity of its contemporary institutions, larger and larger numbers of the population are engaged in and affected by this federal-service element. It is the function of the civil servant to serve the general public by providing services that may offset or combat threats to health or economic stability or that cannot be provided as economically or effectively by an individual acting in his own behalf. The postal system, water supplies, collection and disposal of sewage, aspects of public schools and hospitals, standards of stability in banking and insurance are contemporary areas in which those engaged in civil service endeavor are occupied. A perennial problem in government is how to attain public goals through these services without permitting their abuse for the private gain of factions, individuals, or special interests. This problem is one of the central reasons for the concern and consequent development of techniques for dealing with the classification, recruitment, training, education, lines of career, and pay scales of the civil servant. Developing an individual human being with his needs, desires, and ambitions into a capable and morally responsible agent of the general public is a complex and not always successful task. Commenting on the positive value of a merit system to help achieve this goal, Bernard Rosen advanced six fundamental principles of sound personnel practices, which were set forth in the Intergovernmental Personnel Act:

1. Recruiting, selecting, and advancing employees on the basis of their relative ability, knowledge, and skills.
2. Providing equitable and adequate compensation.
3. Training employees as needed to assure high-quality performance.
4. Retaining employees on the basis of the adequacy of their performance, correcting inadequate performance, and separating employees whose poor performance cannot be corrected.
5. Assuring fair treatment of applicants and employees in all aspects of personnel administration without regard to political affiliation, race, color, national origin, sex, or religious creed and with proper regard for their privacy and constitutional rights as citizens.
6. Assuring that employees are protected against coercion for partisan political purposes and are prohibited from using their official authority for the purpose of interfering with or affecting the result of an election or a nomination for office.[11]

The dual aims of preventing political misuse of administrative positions and of insuring competent personnel through changing political executives are usually enforced in large measure by civil-service commissions or personnel administrators. How to balance the popular consent and confidence from the public with good career and income positions for the civil servant is a perpetual problem for government. Other personnel problems are the tenure of the civil servant; his right to organize to press his group interests in such matters as pay and working conditions — and even policy objectives — against his employer; and the extent of his allegiance to his immediate supervisor, his department head, the departmental personnel office, the chief political executive, the legislature, or the party in power.

Today, the United States Civil Service Commission is no longer merely a national agency but one of international scope. Most military installations in the United States are dependent to a great degree upon civil-service personnel for managerial, skilled, and semi-skilled positions. Partly as a result of this growth, there has been an increasing dependency of society as a whole upon this agency's knowledge, capacity, and integrity. There is a growing affiliation, by law and administration, to intergovernmental functions, and a way of life is developing with its own characteristics shared among civil servants in general and to an intense degree by special groups among them. The term "bureaucracy" perhaps conveys the nature of these societies which exist within the larger society and are organized in a complex of professional and craft associations and other formal and informal relations.

The question now arises as to whether the bureaucracy is serving man or man is serving the bureaucracy. Bureaucracy was described by Honoré de Balzac as "a giant mechanism operated by pygmies."[12] That observation seems to have contemporary validity as modern man diminishes in value before the proliferating process of societal living. Commenting on the same subject, Mary McCarthy declared:

> *The vast growth of the social* [*life*], *steadily encroaching on both private and public life, has produced the eerie phenomenon of mass society, which rules everybody anonymously, just as bureaucracy, the rule of no one, has become the modern form of despotism.*[56]

Civil service, despite its proliferation into an almost unmanageable bureaucratic organism that continues to divide and multiply, has advantages as well as disadvantages for those who serve within it and military personnel who are dependent upon it.

ADVANTAGES AND DISADVANTAGES

The federal government is by far the nation's largest employer. Its payroll covers 2.6 million people (not counting military service personnel), who perform almost every conceivable job. The range of jobs runs through from architect to zoologist and includes beauticians, statisticians, archaeologists, and funeral directors. In years past, the federal government had a reputation as a model employer, but those days are long past. To many federal employees, Uncle Sam appears a "stingy, incurably bureaucratic, highhanded and neglectful boss."[14] Reflections of this kind of discontent are evident from mail strikes and the air-traffic-controller "sick-outs," which represented deep and spreading discontent among federal employees. Noting this discontent, and contending for the right of federal employees to organize and negotiate, one advocate pointed out: "Neither the pillars of city halls nor the foundations of the civil service crumbled when conditions of employment were negotiated instead of being fixed unilaterally."[15]

Increasing numbers of federal employees believe that they are falling behind in wages in the presence of rampant inflation. The new militancy of employees is due in part to the postal workers' illegal strike, but it also mirrors the growing sense of anxiety among blue-collar workers generally. Civil service unquestionably offers a mixture of benefits and frustrations to those working within it.

In the past, one advantage of civil-service employment has been job security, but instances of large cutbacks in employment in the Defense Department and the National Aeronautics and Space Administration (NASA) have rendered this benefit a dubious one. Yet despite such exceptions, job security is underwritten in the civil service merit system. Louise Durbin pointed out that the laws have been particularly beneficial to females in this respect:

> *Once an employee has achieved career status — after three years — she cannot be dismissed without a strong case against her. And she has the right to appeal such action to the* C.S.C. *A federal career can be interrupted without forfeiting benefits accrued. If an employee has career status, then quits, she may return to government work at the same rating she had when she quit, even if it is ten children and 20 years later.*[16]

Other notable benefits are that government employees can eat relatively inexpensive lunches in federal cafeterias, take yearly twenty-six-day vacations after fifteen years and — the biggest benefit of all — retire on full pensions as early as age fifty-five, if they have been employed for thirty years. However, in exchange for these advantages, many federal employees feel that they are circumscribed by inflexible work rules that stifle their initiative and also that they must contend with a rigid salary system that turns ambition to apathy. There are virtually no merit raises, and periodic increases within each job classification are small, employees complain.

In the civil service are also inequities within classifications: whereas the lower-grade classifications (below GS-6) are allegedly paid far less than their skills bring in the private sector; salaries for the so-called supergrades (GS-16 through GS-18) have brought complaints from private employers that they cannot match the federal salary levels for lawyers, economists, and public-relations men. Morale varies widely among departments and agencies and fluctuates with changing circumstances. NASA is an example; the glamor of this agency faded quickly when economy cuts were necessary. In contrast, the Bureau of the Budget has high morale, for many of its jobs are in the higher pay grades, and advancement is quick for those working there. There are other complaints:

> *The least attractive government jobs are in the Social Security Administration, the Veterans Administration and the General Services Administration, which mostly offer jobs in the lower classifications. At the Department of Defense, which employs over 40 percent of all federal workers, civil servants complain that the Air Force hires so many retired officers for top civilian jobs that it cuts off career employees' hopes for advancement.*[17]

The greatest numbers of federal employees are concentrated now, as they have been for the past twenty-five years, in three enormous federal operations: the civilian personnel of the Department of Defense, the postal clerks and carriers of the Postal Service, and the civil servants operating the veterans' services. The civilian components of the armed forces within the Defense Department are the most volatile area of federal employment. Their numbers have increased rapidly in periods of military mobilization and operations — World War II, Korea, and Vietnam — and have decreased with comparable abruptness during demobilization. But it has been observed that, with each:

> *. . . receding tide of employment, the level never quite sinks to what it was before the previous war. The military establishment, with its complex logistical systems and with its extensive involvement in technological research and development, has sustained a high but unstable level of employment. This instability has belied the popular conception that public employment constitutes inviolable security. The shifting of missions, the closing of bases, and the changing of weapons*

systems have forced frequent cutbacks and expansions at a significant number of locations.[18]

Despite such disadvantages to federal employees, many military commanders — far from understanding the problems — add to the predicament by reacting with either indifference or hostility toward civil-service employees. Furthermore, the doctrine that the least government is the best government has tended to color public opinion, even though citizens through their representatives continue to demand more and more services from their government. Both the military supervisor's indifference and civilian hostility are frequently reflected in the popular press. Thus the civil servant is the frequent target of criticism, both serious and humorous. For example, at the time of the launching of the Soviet Union's Sputnik I, the first artificial earth satellite, there was a popular quip that the unsuccessful American space vehicle was named *Civil Service* because "it wouldn't work and you couldn't fire it."[19]

The present demand for high-quality public services cannot be satisfied by demanding a substantial reduction in public employment. In a nation with more than 205 million people with highly diversified lifestyles and with the complexities that have evolved from technological revolution, a large and diversified body of public servants is needed. Also needed are programs to give the public an appreciation of this reality. Finally, citizens must take steps to insure that the federal bureaucracy is a positive and responsive rather than a negative and obstructive force in American life.

Most military supervisors concede that the necessary skills are present within the government to solve the countless problems in the best public interest and to provide the highest talent the nation is capable of developing.

DISCRIMINATION IN CIVIL SERVICE

The federal government's efforts to end discrimination in employment were set in motion by Title VII of the Civil Rights Act of 1964, which forbids discrimination based on race, color, religion, sex, or national origin.[20] The act also established the Equal Opportunity

Commission, which has two primary functions: information gathering and enforcement. The information-gathering function forces the compliance of every enterprise with one hundred or more employees to file annually a form detailing the breakdown by sex and minority race of those employed in each of nine different job categories, from laborers to managers and officials. The enforcement function provides that the commission may sue in a federal district court in its own behalf or on behalf of other claimants when it is believed that discrimination has taken place. The right to go to court thus far has proved beneficial not so much in the courtroom as in the muscle it provides in out-of-court conciliation efforts.

When President Richard M. Nixon signed into law the Equal Employment Opportunity Act of 1972, the United States Civil Service Commission — the federal government's central personnel agency — was given a legislative mandate to see that all personnel actions in the federal government were free from discrimination and that federal personnel management was actively oriented toward equality of opportunity. The act placed federal employees and agencies for the first time under the equal-employment provisions of the Civil Rights Act of 1964, as amended, and placed responsibility for enforcement in the Civil Rights Commission.[21]

Although the preponderance of discrimination complaints originate with blacks and women in the federal civil service, discrimination between upper- and lower-level employees within the civil service itself is worthy of note:

> *Doing the American government's dirty work — cleaning the corridors, filing its memos, carrying its letters — isn't very rewarding, not only because little of the taxpayers' money gets into the payroll for these jobs but, much more regrettably, because personnel treatment at the bottom ranks of U.S. Government is a national disgrace. It makes of those ranks an employment ghetto from which few of the 1.25 million lowest paid workers can ever escape to better positions.*[22]

For some time, low-level civil-service employees have been locked into dead-end jobs where promotion is rarely possible and where there is little training or incentive and few support services. Because the federal agencies had not acted against this state of affairs, Plan D of the Public

Service Careers program (PSC) was started in 1970. This plan addressed itself to the lowest five tiers of federal pay, the 1.25 million workers that the Labor Department considered "low income," using the standard of $6,500 annually for an urban family of four.

Funds were provided to cover part of the cost of training employees on the job and of acquainting supervisors with the problem of discrimination against the low-level employee. The Civil Service Commission identified supervisor awareness as a key element that should underscore the important role the supervisor plays in the successful adjustment of the public-service career trainee in the work situation. Approximately $10 million has been spent on these programs in federal agencies since the inception of Plan D involving the commitment to train about 18,000 employees to rise in the ranks of federal civil-service employment. Commenting on some lack of cooperation in this venture, one production-line supervisor pointed out:

> *Directors, administrators and others at the managerial levels of government have abundant opportunity to attend lunch-and-cocktails sessions, meetings, seminars, and lectures, not to mention professional training programs, but the low-pay messenger is "so important" that "he can't be spared to see a movie" or spend an hour or so a week in a course designed to give him a new skill.*[23]

In an attempt to alleviate discrimination against lower-level civil-service employees, several Washington, D.C., employees trying to make a "sensitivity" film to tell their story claimed censorship of their efforts by some members of the Civil Service Commission. Mrs. Tina Hobson, director of Agency Operations within the Public Service Careers Office of the Civil Service Commission Bureau of Training, cited the difference between the audience to whom the film should be presented and those low-level civil-service employees who acted in the film to depict the kinds of discrimination from which they suffered. The typical audience was made up of management-level decision makers, all white, almost all male, and almost all older than forty-five. One of the problems cited was how distant the decision-makers were from the problems and the employees who were most directly affected by them — most of whom were nonwhite, female, and young.

Those involved in producing the film were cautioned against making the film an "equal-opportunity" presentation, which was interpreted to mean: don't concentrate on blacks, include Indians, Spanish-speaking Americans, and white women, because they, too, are victims of job discrimination. The film makers countered by pointing out that in Washington, D.C., the problem is in fact almost invariably black-white. Those responsible for making the sensitivity film also declared:

> *We were warned against using any statements that reflected badly on any federal agency. Other statements that had to be censored from the film usually bore too heavily on the racial aspects of the oppression suffered by low-level federal employees, but sometimes remarks were cut simply because they were too caustic about the Establishment, the system, the bureaucracy.*[24]

Those involved with the film and its effort to raise the spirits of the lower-level workers bitterly concluded that, as far as the Civil Service Commission was concerned, executive training had top priority and that job and skill training and upward mobility for the minimally skilled had low priority.

RACIAL DISCRIMINATION

When the United States government was formed, it was generally assumed that blacks would not be employed in official federal positions, and by the end of the 1820s, Congress had specifically prohibited the employment of blacks in the Post Office. The Civil War did little to improve the opportunities available to black Americans in government service, though as a token gesture a few blacks were given posts in Haiti, Liberia, Santo Domingo, and the District of Columbia. Not until the passage of the Ramspeck Act and the accompanying Executive Order of November 7, 1940, was racial discrimination made illegal in federal employment and promotion policies.[25]

Since the mid-1950s, the federal government has taken a number of major steps to make certain that racial discrimination in civil-service employment is eradicated. Nevertheless, one of the major concerns of minority groups is the Federal Service Entrance Examination, which is often cited as reflecting cultural bias at its worst. In 1967, a survey of 191 colleges and universities showed that, while 71.7 percent of those taking the examination passed it, only 29.6 percent of the students at predominantly black colleges who took the examination passed it.[26] The tests were predicated on the assumption that the person with the highest skills or the greatest educational ability would be the best performer. In reality, the test was largely a measurement of white college graduates with reasonably similar educational and cultural backgrounds.[27]

Ollie A. Jensen stated: "Often the level of verbal ability needed to pass a civil service test is far in excess of that required to be successful on the job. This needlessly discriminates against competitors with particular ethnic or socio-economic backgrounds."[28] A number of experts have pointed out that questions that directly relate to the skill desired in some concrete way should make up most of the items on the examination. Questions that require a knowledge of language unrelated to the particular job should be eliminated. In the long run, the patterns of bias in the Federal Service Entrance Examination can only be destructive to the goals of democracy:

> *Expediency and tight-fisted economy in the operation of a selection program tend to produce the same result as corruption and incompetence — a downgrading of one segment of the competitive group and the upgrading of another segment for invalid reasons.*[29]

The administrations of Presidents John F. Kennedy and Lyndon B. Johnson brought some remarkable gains for blacks in civil-service employment. Under President Kennedy, the government set about recruiting black students on college campuses for civil-service employment, and many of these students were the beneficiaries of counseling programs relating specifically to federal employment. Under President Johnson, the Civil Service Commission set up a program to eliminate discrimination based on four objectives: (1) a determination to eliminate prejudice at all levels of federal service; (2) a thorough reappraisal of job structure and employment practices; (3) a strong emphasis on training and upgrading employees already on the rolls; and (4) a strong campaign for community involvement.

Mexican-Americans, like black Americans, have for many years labored under the burden of racial discrimination in federal employment. During the 1960s, the Mexican-American employees of Kelly Air Force Base in San Antonio began to protest discrimination at that installation. The United States Civil Rights Commission began hearings to determine the causes of the dissatisfaction and found that, though Mexican-Americans made up a sizeable segment of the work force at the base, they were all too often unable to achieve advancement or to reach supervisory or other senior positions in management. A second complaint by Mexican-Americans at the base was that Anglos failed to learn Spanish while non-Anglos were forced to learn English. John W. Macy Jr., described the situation at Kelly Air Force Base by saying, "The development of bilingual and bicultural objectives had not been accepted by the leadership in the broader community and consequently was resisted in the work situation at the base."[30]

In 1968, Julius W. Hobson, a black economist with the Social Security Administration, charged that "Uncle Sam is a bigot." To support this claim he submitted a number of facts: before 1968 the Federal Communication Commission had never hired a black programmer; although 320,136 of the 2,304,000 classified civil servants were black, 88 percent of this minority group were in the lowest-paying jobs; nine out of ten black federal employees in the District of Columbia were in the bottom eight grades of the civil service, and half of these were in the bottom four grades.[31]

Hobson's charges seemed to be substantiated to some extent by the fact that, in 1970-1971, while blacks made up 22.3 percent of all employees in General Schedule grades 1 to 4, only 1.6 percent of those in grades 16 and above were black.[32]

Nevertheless, minority employment and status in the federal civil service appear to have improved to a significant extent in the past few years. As of May 31, 1973, minority-group members — Oriental-Americans, Indians, blacks, and Mexican-Americans — held 515,129 federal jobs, or 20.4 percent of the total. Minority groups made up 28.3 percent of the total work force in Grades 1 through 4; 20.5 percent of the total work force in Grades 5 through 8; 10 percent of the total work force in Grades 9 through 11; 5.8 percent of the total work force in Grades 12 and 13; 5 percent of the total work force in Grades 14 and 15; and 3.5 percent of the total work force in Grades 15 through 18. The largest percentage gain during the period 1972 to 1973 was shown by Spanish-surnamed employees, who accounted for 3.1 percent of total federal employment.[33] Between 1972 and 1973 the percentage of minorities in the better-paying white-collar jobs rose from 15.5 to 16.3 percent.[34]

SEX DISCRIMINATION

Soon after John F. Kennedy took office as president in 1961, he turned his attention to equal opportunity for women in federal employment. Yet it was not until after the beginning of 1964 that serious, concrete steps were taken to mark the commencement of a program to root out sex discrimination in the government. In 1963, President Johnson ordered a talent search for women, and between January 1, 1964, and October, 1965, federal agencies appointed or promoted more than 3,500 women to positions at salaries of $10,600 or more. When he received the winners of the Public Service Career Awards at the White House in the spring of 1965, President Johnson noted that all those present were male and severely criticized the selection committee for failing to select women.

On May 12, 1971, the Civil Service Commission ruled that federal agencies could not prescribe "men only" or "women only" jobs, a major step forward in federal employment. In 1969, it had been found by the Civil Service Commission that 50 percent of the employees in GS levels 1 through 6 were women; in GS 7 through 12, only 20 percent were women; in GS 13 through 15, only 4 percent were women; and at GS 16 and above, only 1.7 percent were women. Of all women holding federal jobs, 78.1 percent worked in GS 1 through 6 positions; 20.8 percent, in GS 7 through 12 positions; 1 percent, in GS 13 through 15 jobs, and in GS 16 and above, only .03 percent.[35] It is clear that the federal government has a long way to go in abolishing sex discrimination in its hiring and promotion practices.

A Federal Equal-Employment-Opportunity Program

Major Gary R. Lee,
Secretary of the General Staff
U.S. Army Troop Support Command, Missouri

BACKGROUND

The U.S. Army Troop Support Command (TROSCOM), with headquarters in St. Louis, Missouri, is one of the seven major logistical commands responsible for the wholesale supply and maintenance of the materiel needs of the U.S. Army and, in some instances, gives support to its sister services throughout the Department of Defense. Specifically, TROSCOM is responsible for the integrated materiel management, production, maintenance engineering, research, design and development of materiel, and procurement of assigned categories of equipment ranging from engineer construction equipment and bridging, the Army marine fleet and rail equipment to such items as clothing, sentry dogs, food, camouflage, and topographic equipment. The command employs approximately 5,600 personnel, of which about 3,200 are assigned to the headquarters in St. Louis. Significantly, the headquarters is located within the city limits of St. Louis.

Disciplines represented by the command's employees range from some of the leading research scientists in the country to custodial maintenance. The command has a total budget in excess of $350 million and a $240-million procurement program. The bulk of the command's personnel not located in St. Louis is a part of the research, development, and engineering activity, located at the Mobility Equipment Research and Development Center (MERDC) at Fort Belvoir, Virginia, with a budget of approximately $52 million a year, and Natick Laboratories in Natick, Massachusetts, with a budget of approximately $32 million a year.

Although the military services have been in the vanguard of organizations achieving first integration and subsequently progressive equal-employment-opportunity (EEO) programs, their most notable achievements were largely confined to their military personnel, their civil-service work forces lagging appreciably behind. Thus it was not surprising to find that, in the study of the EEO programs up to and including 1968, the Army's logistical commands with primary civilian work forces were still struggling to provide equal opportunity for minority groups and women employees. However, even within the study of the profile of a seemingly successful program, much is still required to be done.

Research into this program indicated that a higher level of directives, that is, executive orders from the president, guidance from special assistants to the president, directives from the Department of the Army, and so on are most important in providing overall program guidance. They lacked, however, the specificity necessary to make the program a success, although they did give important impetus to the commanding general's request for results. The result was a noticeable decline in direct reference to executive orders, Congressional actions, and so on, although it should be recognized they did set the all-important stage for affirmative actions, the enigmatic nature of interaction with minority groups and women and the inevitability of certain changes eventually to come.

Following is a chronology of those decisions that developed and sustained a positive program at TROSCOM and eventually ideas essential to further understanding needed for the change.

PROGRAM ACTIONS

Initially one overriding decision had to be made and accepted at every executive level of management within this large, complex organization before real progress could be made. This decision, and employee recognition, were not initially published or officially recognized, and yet they really provided the spark to achieving acceptable results and set the stage for later decisions and actions that moved the command EEO program forward in specific areas. The decision was that, although we were implementing an EEO program, the façade of equal opportunity needed to be abandoned within the executive decision-making stratum. It had to be recognized that mere equal opportunity to compete for a position or for training would not produce the desired results. The periods of time in which the results were desired were such that in many cases, minority-group employees who were expected to compete were not in fact equal in previous training, prior experience, social background, motivation, competing grades, and so on. Therefore, special consideration needed to be given to these employees; that is, at a minimum, a weighing factor had to be added to employee evaluation to insure truly equal competitors. This consideration would usually lead to numerical goals that allowed minority groups to compete for a fair share of opportunities.

Since it was an extremely difficult transition for many managers to make and accept, extremely strong command guidance was required on an almost individual basis to assure at least overt acceptance. Command-wide acceptance at the executive level of this initial premise was not, of course, accomplished simultaneously with the recognition of needs. Top management seemingly accepted this premise in late 1968 and early 1969, but it was not fully accepted by the remaining executive stratum of TROSCOM's management even as late as the latter part of 1970. Even later, representation could be noted of the minority of women employees selected for long-term training, logistical and procurement intern candidates. However, later breakthroughs in training, both long term and short term, began increasing the percentage of minorities and women employed.

The next decision to have a major effect on the EEO program at TROSCOM was the decision to appoint an individual to devote full-time to the program. This command was the second organization in the St. Louis metropolitan area to have a full-time EEO program manager. This decision enabled the command to have the necessary continuity basic to its success. It allowed an individual outside the personnel and training directorate to address himself to minority group and women problems separately and objectively. It enabled, for example, independent surveys to be conducted of the command to determine the existence of factors detrimental to program progress. This individual could assist managers in determining causative factors of the symptoms discerned during the surveys. These surveys were basic in that they both assisted the lower-level supervisors in identifying and addressing themselves to the problems in their area and provided data for top management. When short-range problems were found, target dates for their solution were established and they became formal agenda items for weekly staff meetings conducted by the commander and key staff members. Long-range plans and problems were the responsibility of those involved in implementing the affirmative action plan.

Another key decision that dovetailed with the establishment of a full-time, equal-employment-opportunity officer was that of the commander, who required a formalized review of promotion actions that affected women and minority-group employees. This review, assigned to the EEO manager, resulted in better justification for selection in all promotion actions. The review of promotions significantly enhanced the upward mobility of women and minority-group employees. For example, from July 1, 1968, the average minority grade increased steadily from GS 5.3 to GS 6.848 on the General Schedule (GS) salary scale, whereas the overall command average, including minority-group employees, rose only 0.8 percent from 7.6 to 8.4. In addition, as of July 1, 1968, only 19.3 percent of minority-group employees occupied positions as GS 7 and above. Within a short time, 43.1 percent of the minority-group employees on the payrolls were in grades GS 7 to GS 14. This compares with an overall average of 67.7 percent of all employees above GS 7 level. Also, minority-group employees moved into the GS 13 grade level and represented 4.6 percent of the employees at that level. Although these increases are significant, they were not felt to be adequate, and as a result, the command established numerical goals to assure an increased rate through upward mobility.

In late 1968, the command significantly strengthened its internal system for processing minority-group complaints with the objective of having the internal system so responsive that few complaints would have to be handled by a higher headquarters. This system has been gradually strengthened and has been extremely successful. Only very rarely are internal actions seriously questioned by higher headquarters upon appeal.

The command EEO officer is provided assistance by selected area EEO counselors from among the employees throughout the command. It has been found that minority-group complaints can be resolved by the complainant working out an adequate solution with the area EEO representative. Strength in the command EEO program has been recognized through the efforts and confidence now formalized by management through the area EEO counselors. An area EEO counselor is selected by the specific management area from among the employees and coordinates with the command EEO officer to assist in reviews, investigations, and the eventual satisfactory resolution of minority group complaints.

A command decision which holds much promise was the establishment of a full-time EEO officer at TROSCOM's subordinate command, MERDC, at Fort Belvoir, Virginia. The problems were intensified there because the disciplines in that organization were and are almost entirely scientific and engineering. Much effort remains in this area both in recruitment and in training. They did, however, achieve progress over a period of two to three years from about 5.4 to 9.3 percent minority-group employees.

In late 1969, the commander directed a review and study of the command's training and awards program, one of the objectives being to discern the impact of these programs on women and minority-group employees. The results of this survey revealed that only 20 of the 355 employees who received off-post training during the review period were women and minority-group employees, who represented less than 6 percent of the total who received such training. Also the minority-group employees, who accounted at the time for approximately 15 percent of the total work force, only received about 12 percent of the total on-post, noncollege training provided by the command and had received no allocations for "long term," that is, full-time, college-level or postgraduate training.

Because the area of training is extremely important from a developmental standpoint, the action taken to rectify the situation was immediately to establish numerical goals to bring the level of training of the women and minority-group work forces up to that of the remainder of the command. This effort was successful in that the 6 percent was increased to 33 percent and the 12 percent for on-post training was raised to parity with the percentage of minority group employees in the work force. In 1973, minority employees accounted for 50 percent to 51 percent of those receiving long-term college-level training.

In the area of awards, a survey of the period 1969 to 1973 indicated that the minority-group employees received only 10 percent of all awards. Although no quotas were established in this area, reviews were conducted of each of the major organizational elements at the weekly staff meetings, giving the commander an opportunity to comment and act on the manager's progress in this area. As a result of the commander's increased awareness of this management technique, the minority-group employees were soon receiving approximately 20 percent of all awards.

In April, 1973, the Incentive Awards Committee and Training Committee added the command EEO officers to their memberships. This has added the much needed impetus to insure proper distribution and adequate representation for selection to long-term training courses and government training courses and proper distribution of awards to minority-group employees.

During this period, the command underwent a substantial reduction-in-force, but, as a result of special consideration and review of minority-group employee actions, care was exercised to insure truly equal consideration of reduction actions that affected these employees. This was one of the major contributing factors to the fact that after the reduction, minority-group members were approximately 17.4 percent of the work force. This percentage exceeded the minority representation in the St. Louis metropolitan area, which was about 16.4 percent. These changes in minority-group representation were considered quite an achievement, inasmuch as TROSCOM experienced a reduction-in-force in 1970 — and, owing to a reduction in personnel in the armed forces, 1180 persons were eliminated in 1973 and another 400 were eliminated in 1974, primarily through transfer of a mission to another command.

The command has initiated a strenuous effort to increase its Domestic Action Program, wherein the management and labor forces attempt to overcome some of their serious local domestic problems. Communications with local leaders have been expanded, and command assistance to local youth groups and other organizations has expanded as rapidly as concerned individual employees could be identified who would volunteer to assist during their off-duty hours. The command's Domestic Action Council sponsors a number of projects that materially benefit the youth from the minority community. These include the sponsorship of five Junior Achievement Companies and several Explorer Scout posts (predominantly black), and support for the St. Louis Children's Study Home and the Annie Malone Orphan's Home. The command EEO served as project officer for the study home. Support provided in this connection was sufficient to merit a letter from the mayor of St. Louis, to the command. Junior Achievement fairs held within the facilities of TROSCOM for the Junior Achievement Companies netted sales of several thousand dollars.

FEDERAL WOMEN'S PROGRAM — TROSCOM

Presidential Executive Order No. 11375 was signed on October 13, 1967, by President Lyndon B. Johnson.

> *It is the written policy of the Government of the United States to provide equal opportunity in Federal employment for all qualified persons, to prohibit discrimination in employment because of race, color, religion, sex, or national origin, and to promote a full realization of equal employment through a positive, continuing program in each executive department and agency. The policy of equal opportunity applies to every aspect of federal employment, policy, and practice.*

To insure the implementation of the order, the Civil Service Commission directed the establishment of the Federal Women's Program (FWP). The position of a program manager was established within the Civil Service Commission to be the overall coordinator of the FWP. The *Federal Personnel Manual (FPM) Letter 713-8,* dated January 25, 1968, became the document by which the Civil Service Commission began its EEO program for women, under the overall coordination of the program manager, the FWP coordinator. This letter placed requirements on all government agencies to implement similar programs to overcome known discriminatory practices against women. The letter was the amendment reinforcing existing programs and strengthening what the EEO program had already been established to accomplish. The key to remember here is that the letter was to give strength to the already-existing program within EEO.

Then the Department of the Army, under Civilian Personnel Regulation 700 (Change 13, Subchapter 2, Section 713.2, dated November 10, 1972), identified the parameters for organizing the FWP within the U.S. Army. The CPR not only outlines the whole of the EEO program but addresses and spells out the duties and responsibilities of the FWP coordinator and the implementation of the program, its committees, and its responsibilities at the government activity level. This document also includes the FWP coordinator's primary duties. The military has been slow to establish a meaningful and recognizable FWP, mainly because it is a part of the overall EEO program. Therefore, most military installations have a part-time coordinator as at TROSCOM.

TROSCOM has long sought relief from the barriers that have prevented many high-potential employees occupying low-graded, dead-end positions from entering Army career programs that offer greater promotional opportunity, as well as the potential for greater job satisfaction. However, efforts toward this end have been frustrated because of the conditions, requirements, and controls imposed by higher departmental levels. Military commands have had little control over input (interns) into Army civilian career programs. TROSCOM's own initiative, through establishment of its Upward Mobility Program, has begun to satisfy this need.

The Upward Mobility Program outlines proposed actions that would facilitate entry of employees into various stimulating career programs. A requirement was placed upon all major staff offices to develop realistic upward mobility programs for their underutilized, high-potential employees with strong emphasis placed on the creation of better job opportunities for many employees limited to dead-end, clerical positions.

TROSCOM's approach has been to provide needed training and development for women. The first approach was to provide a means by which women could acquire much needed skills and training. We have had tremendous results in this early part of TROSCOM's approach to improving skills and training. The command participated in a seminar at the University of Missouri at St. Louis (UMSL) on an "Affirmative Action Program for Women." The seminar was primarily addressed to the business community; however, the groundwork was for joint UMSL-TROSCOM-EEO efforts to design a program that could benefit the women of TROSCOM, as well as those elsewhere employed within the federal community. As a result, the St. Louis Federal Women's Program coordinator, the EEO officer, UMSL, and top management aim toward developing a more effective and productive FWP for TROSCOM.

The UMSL designed two special eight-week courses specifically requested by the women employees of the command. One was "Advanced Professional Development secretarial (girl Friday) Course," and the other was "Management for Women Course." Both courses were conducted once a week during after-duty hours by highly skilled, well-qualified faculty members of UMSL. These two courses were well-received and have been offered several times. No restrictions were placed on individuals who could attend the courses, women or men. Each of the graduates was fully reimbursed by the command for cost of fees, books, and supplies.

Later the women, in conjunction with UMSL, arranged and provided sponsorship of the College Level Examination Program (CLEP). Approximately thirty individuals completed the entire program, of whom twenty-six were women. Additionally, arrangements were made for vocational testing and counseling to be provided TROSCOM through the university program.

Because of the success of the first two courses and the vocational testing and counseling program for women striving for upward mobility, another course. "Assertiveness Training for Women," was created. This course is designed to develop a woman's potential for asserting herself in a new role as opposed to the traditional stereotyped roles.

A program in vocational counseling, "Self-Directed Search," was attended by 380 employees, eight of whom were men supervisors. A total of 316 completed the program. It cost each individual only three dollars, and was a most helpful and educational program to highlight the efforts of the FWP committee. The committee meets periodically and works very closely with UMSL. At several of their meetings, the question of day-care center services for TROSCOM parents was consistently one of the FWP committee's concerns, and as a result the UMSL opened a day-care center near the campus. The center was opened not only to the students on the campus but to the community as a whole. And because of the support of TROSCOM, FWP committee, the university provided access to its facilities to TROSCOM parents. The day-care center idea was initiated and originated through questionnaires distributed to TROSCOM personnel. The university looked toward TROSCOM parents, who were deeply interested, to insure that the center provided what parents wanted and needed, and, of course, to allow women the training that is provided and to work toward higher-paid skills. At this writing, the committee continues working to determine other areas of interest for the women within the community of TROSCOM.

In the meantime, Federally Employed Women (FEW), a national organization, was making plans to organize a FEW chapter within the St. Louis metropolitan area. As a result of TROSCOM's FWP organization, FEW solicited support and assistance of our EEO officer and FWP committee and established a subchapter of FEW within the St. Louis federal community.

As a result of these activities, the Department of the Army presented its second-highest award, the Meritorious Civilian Award, to the TROSCOM Equal-Employment-Opportunity Officer. This action was a first for two reasons: The EEO officer was male and black, and he had been nominated by a committee of TROSCOM's FWP. It was the committee's findings that he had been most cooperative and most instrumental in furthering the efforts of equal employment for women of the federal community. Later he was invited to become an honorary member of the FEW.

So I can say without doubt that the FWP at TROSCOM contains a large number of active women, full of tremendous ideas and ambitions to break out of their traditional role and become recognized as human beings. They have been most successful in their endeavors and are recognized in the metropolitan area with an active plan and — of most significance — have positive results.

FUTURE PROGRAM DIRECTION

The principal areas of emphasis, planned to assure a continued affirmative program, involve, first, continuing to stress upward mobility until parity in relative grade level is achieved. The numerical-goal concept as opposed to strict quotas is felt to be the best path leading to the accomplishment of the objective, along with the expected benefits derived from training efforts, particularly in long-term training. Additional emphasis is being applied to breaking up concentrations of women and minority-group employees and insuring that they are dispersed throughout all career fields, scientific and engineering disciplines, as well as administrative and logistical support areas. This will be accomplished principally through local surveys to determine racial concentration and in-depth counseling to ascertain employee attitude and interests. There will be a relaxation of emphasis in several of the areas where program successes have led to parity with the overall work force.

Another important goal recently achieved has been the modification of the command's automated data system to insure that the necessary data on women and minority-group employees are available to assist in surveys and studies planned for the future. These data will provide a most useful managerial tool if used properly and purposefully. Finally, a greater degree of program decentralization will be attempted as acceptance of the program is assured at increasingly lower levels of management.

The success of the U.S. Army Troop Support Command's program has been primarily due to foresight and diligent effort on the part of its executive management. The above decisions in no way present the total effect of the program's effort but summarize the key decisions that have yielded the greatest benefits and assured the program's level of success. It is worthy of note that in virtually every case, the command's management made these and other decisions before receiving direction from higher command headquarters or other federal agencies. The decisions were made on the basis of an internal determination of the need to meet the overall governmental objectives.

Although detailed statistics certainly attest to a successful program, perhaps more important is the readily discernible attitude on the part of most of the command's employees to strive to make the program a success, not only from a statistical standpoint but also from the standpoint of a truly harmonious, cohesive, and satisfied work force. The EEO program has earned the respect and confidence of women and minority-group employees, as well as top-level management. A combined effort of all employees to work out solutions to problems and the ability to lay everything on the table and make positive communication with each other has further led to the successful, but still growing Upward Mobility Program and Affirmative Action Plans. Progress is apparent in most cases, and the employees are learning to know and understand each other, and how to care for people, and how to recognize each other as responsible human beings without prejudices.

This nation is a single, unified, political and geographical entity inhabited by human beings who represent many and divergent nationalities, races, religions, and cultures. This nation is the land of opportunity and the home of the free, but many groups have become alienated. Political and economic colonialism in this country has had social repercussions dating back to slavery and did not end by any means either with the Emancipation Proclamation or with the renaissance of the women's movement.

The Federal Equal Employment Opportunity Program at TROSCOM has changed the relationship of human beings, but I still see a long road ahead. The establishment of strong and positive programs has required hard work within the command, but total commitment is required of all employees to correct the many imbalances of long standing within the female and minority-group work forces.

Continued emphasis on the development of a strong Equal Employment Opportunity Program stands high on the list of priorities throughout TROSCOM because:

1. First and foremost, it is the right thing to do.
2. It can provide effective utilization of the total capacity of the work force.
3. It is national policy.

The adaption to change rests on how effectively we — whites, blacks, orientals, Indians, men and women — can understand, channel, and direct our energies. These energies must maximize the opportunities for our development and maximize each individual's opportunities to approach the utmost expression of humanity of which he or she is capable.

Implementing a Plan

Lt. Commander W. K. Wible,
Officer in Charge
Navy Telecommunications Center, Washington, D.C.

BASIS FOR INTERVENTION

U.S. Naval Support Activity Instruction 5350.1
From: Commanding Officer 30 November 1973
Subj: Human Relations Council; establishment of
Ref: (a) OPNAVINST 5350.1 of 17 September 1973

1. *Purpose.* Reference (a) directs that each Naval Commander establish a Human Relations Council in order to provide a formal method for maintaining a channel of communication regarding human relations matters within the command.

2. *Discussion.* Naval Support Activity, Canal Zone, personnel consists of 16 officers, 47 enlisted, 265 full-time permanent, appropriated, funded civilian personnel, and about 115 nonappropriated, funded employees. Employees are U.S. and Panamanian citizens, men and women, of black, white, yellow and red races, and of various national or ethnic origins including Asia, South America, Europe, West Indies, Africa, etc. Military personnel assigned are U.S. citizens belonging to various ethnic groups. Many religions are represented. In view of the multi-racial/ethnic makeup of personnel assigned, there is a potential

for human conflict and tension based on actual or imaginary acts of discrimination or prejudice. The Human Relations Council is established in order to provide a forum for surfacing and resolving grievances related to discriminatory action or prejudicial attitudes.

3. *Organization.* The Human Relations Council membership will consist of the following:

Executive Officer (Chairman)
Civilian Personnel Officer (Vice-chairman)
Union Steward(s)
Equal Opportunity Counselor(s)
Senior Enlisted Advisor
Chaplain
Enlisted Minority Member(s)

Normally the Human Relations Council will meet on the third Thursday of each month.

4. *Action.* The Human Relations Council will support the chain of command providing a means of improved communication and awareness. The council will accomplish this by:

a.) Providing a forum for the surfacing and frank discussion of real or imagined grievances that cut across division or departmental lines. Generally, internal division or departmental problems should be handled within the division and through the chain of command.

b.) Developing and sharing ideas that will foster harmonious human relations, periodically scheduling meetings with civilian, enlisted, and officer personnel to seek their input.

c.) Collecting data about incidents and situations where tensions, dissension, or discrimination may exist and proposing corrective action; following up by soliciting constructive feedback to ascertain whether climate has improved.

d.) Assisting the commanding officer by insuring that the policies and programs of the command are widely disseminated and known at the lowest level.

e.) Advising the commanding officer on the effectiveness of the command's human relations and possible methods of enhancing its effectiveness.

f.) Informally monitoring the effectiveness of the command's equal opportunity program and advising the commanding officer of areas needing improvement.

g.) Promoting intercultural and interracial understanding by providing for a free interchange of ideas between members of different ethnic and racial groups.

GETTING STARTED

From: Chairman, Human Relations Council 20 December 1973
To: Commanding Officer, U.S. Naval Support Activity, Canal Zone
Subj: Human Relations Council Meeting,; minutes of
Ref: (a) U.S. NAVSUPPACT INST 5350.1 of 30 Nov. 1973

1. In accordance with reference (a), the Human Relations Council met on 20 December 1973 at 0900. The following people were present:

Executive Officer
Local 907 Steward
Equal Opportunity Counselor
Chaplain
Senior Enlisted Advisor
Minority Affairs PO

2. The following items were discussed:

a.) Communications gap between management and employees in the Navy Exchange, particularly in the warehouse, causing mistrust and fear of reprisal. Two actions were considered:

(1) To have a meeting with Navy Exchange Officer, Executive Officer, Equal Opportunity Counselor, and Local 907 Steward to further discuss the problem.

(2) Conduct orientation meetings with employees to explain employee management relationships, duties, responsibilities, and privileges of employees. Also discuss the reasons behind the management procedure changes which have occurred over the past year.

b.) It was noted that the enlisted men assigned to the waterfront desire additional training in technical areas. This was brought up in recent Upward Seminars. The Executive Officer is therefore investigating instruction in diesel mechanics, air conditioning/refrigeration, and electrical systems.

c.) It was suggested that a member of FEW, be invited to be a member of this council.

d.) Conflict between Navy Exchange Anchorage Club Hot Dog Wagon and personnel in Barracks No. 66. Conflict included complaints of personal abuse. This matter was referred to two Petty Officers for resolution. It was noted that abusive language was one cause behind this problem. It is not to be tolerated.

3. The meeting adjourned at 1000. The next meeting will be held at 0900 on 17 January 1973 in the Executive Officer's office.

THE TIP OF THE ICEBERG

The Naval Support Activity Human Relations Council is presided over by the Activity Executive Officer. Membership includes those representing minorities, the Equal Opportunity Staff, Drug and Alcohol Abuse Counselor, Director Civilian Personnel and the station Chaplain. This group has been meeting monthly since December, 1973. A list of topics discussed is indicative of the general concerns that prevail in the activity.

1. Abusive language towards black Panamanian by black American sailors.
2. Lack of training (vocational type) available to nonrated sailors.
3. Lack of communications between management and employees of the Naval Exchange system.
4. Need to conduct orientation meetings with employees to discuss employee-management relationship, duties, responsibilities, and privileges.
5. Evaluation — why blue-collar workers get fewer outstanding evaluations as compared to white-collar workers.
6. Time Clocks — why time-clock policies seem to be inconsistent.

7. Promotion procedures at the Navy Exchange.
8. Special recognition for outstanding sailors.
9. Need for language training (in English) to improve civilian promotion opportunities.
10. Poorly maintained restroom facilities for women.
11. Poor working conditions of the Navy Exchange Garage, lack of coffee breaks, too short lunch breaks.
12. Lack of orientation for newly arrived civilian employees and sailors.

Some of these problems are recurring but as a result of management being made more acutely aware of items listed above, the following has been accomplished.

A new standardized plan is being worked out to eliminate time-clock punching requirements for all supervisors and all nonsupervisors in or above GS-6/MG-9 levels throughout all Naval facilities in the Canal Zone.

The Civilian Personnel Officer and Employee Relations Officer have briefed all supervisors on the evaluation procedures and made themselves available to assist by writing justification statements for blue-collar supervisors desiring to give outstanding evaluations to persons under their supervision.

Although not directly related to the existence of the Human Relations Council, the command has transferred the Navy Exchange personnel administrative function from the Exchange Officer to the Director of Civilian Personnel in order to provide some consistency in hiring procedures, reduction in force actions, evaluations, etc.

The president of the local union has been invited to meet regularly with the director of civilian personnel to discuss matters of common interest.

The training of nonrated enlisted men has become more systematized.

The station chaplain is a member of the council and serves as the Human Relations Council chairman's confidant and assists in evaluating the accomplishments and effectiveness of the council.

The apparent problems that were easily correctable have been corrected. Specific complaints are being processed through the established Civil Service appeal system. A persistent problem is to avoid loss of interest because of stagnation and apathy. However, moving too fast in this atmosphere may damage any progress already made. A credibility question frequently raised by the members of the council is: Why meet just to talk?

The Human Relations Council does not have full support of all supervisors. The council is not an investigative body. Its purpose is to serve as a communications link between various groups of the command and the Commanding Officer's representative.

QUESTIONS FOR FURTHER DISCUSSION

1. What techniques could be employed to instill *esprit de corps* in civil-service employees in the lower levels?
2. Should skills proficiency be a part of the Federal Service Entrance Examination?
3. Should more emphasis be placed on human relations workshops to avoid a repetition of incidents like those at Kelly Air Force Base in the 1960s?
4. Should there be a review of classifications for civil-service personnel every two years?
5. What are the best means that can be used to acquaint the public with problems confronting civil service employees?
6. What incentive factors can be offered to insure high-quality employees for overseas duty?
7. Should military personnel be assigned to all jobs currently held by civil servants in military installations?

Notes

PART 1

[1]Friederich Hölderlin, quoted in Octavio Paz, *The Labyrinth of Solitude* (New York: Grove Press, 1961), p. 26.

[2]Rabindranath Tagore, *The Collected Poems and Plays of Tagore* (New York: Macmillan, 1913), p. 245

[3]See Robert Tannenbaum, Irving R. Weschler, and Fred Massarik, *Leadership and Organization* (New York: McGraw-Hill, 1961).

[4]Ralph Waldo Emerson, *Conduct of Life* (Boston: Ticknar and Fields, 1860), p. 42.

[5]Bernard M. Bass, *Leadership, Psychology and Organizational Behavior* (New York: Harper & Row, 1960).

[6]Fred E. Fiedler, *A Theory of Leadership Effectiveness* (New York: McGraw-Hill, 1967).

[7]Donald C. Pelz, "Influence: A Key to Effective Leadership in the First Line Supervisor," *Personnel,* **29** (November, 1952), 209-17.

[8]Ralph M. Stogdill, "Personality Factors Associated with Leadership," *Journal of Psychology,* **25** (1948), 35-71.

[9]Robert Tannenbaum and W.H. Schmidt, "How to Choose a

Leadership Pattern," *Harvard Business Review,* **36** (March, 1958), 95-101.

[10]Ramón Lopez, *Overseas Weekly,* December 3, 1973.

[11]Napoleon Bonaparte, quoted in Emerson, *op. cit.,* p. 40.

[12]Jacob Bronowski, quoted in Rollo May, *Power and Innocence* (New York: W.W. Norton, 1972), p. 165.

[13]Joy P. Guilford, *General Psychology* (New York: D. Van Nostrand, 1952), p. 121.

[14]Adolph Hitler, quoted by Dorothy Thompson in the *New York Post,* January 3, 1944.

[15]Abraham Lincoln, Annual Message to Congress, December 1, 1862.

[16]Kaiser Wilhelm II, speech to his recruits, 1891.

[17]Emerson, *op. cit.,* p. 39.

[18]*Ibid.*

[19]Oscar Wilde, quoted in George Seldes, ed., *The Great Quotations,* (New York: Pocket Books, 1960), p. 728.

[20]Ariwara Yukihira, *Japanese Verse* (Baltimore: Penguin Books, 1964), p. 78.

[21]Francis Bacon, *Novum Organum,* ed. by Thomas Fowler (Oxford: Clarendon Press, 1889), p. XIX.

[22]Albert Einstein, "Religion and Science," *New York Times Magazine,* November 9, 1930, p. 1.

[23]Emerson, *op. cit.,* p. 49.

[24]Ralph Waldo Emerson, "Ode," 1847.

[25]Kurt Lewin, Ronald Lippitt, and R.K. White, "Patterns of Aggressive Behavior in Social Climates," *Journal of Social Psychology,* **10** (May, 1939), 271-99.

[26]Lester Coch and John R.P. French, Jr., "Overcoming Resistance to Change," *Human Relations,* **1** (1948), 512-32.

[27]Lao Tzu, *The Way of Life According to Lao Tzu, An American Version* by Witter Bynner (New York: John Day, 1944), pp. 34-35.

[28]Fiedler, *op. cit.*

[29]Edwin P. Hollander, *Leaders, Groups and Influence* (New York: Oxford University Press), 1964.

[30]Edward L. Benays, "Put Your Idea into Action," *Freedom and*

Union, October, 1947, p. 20.

[31]Thomas Wolfe, "The Anatomy of Loneliness," *American Mercury,* October, 1941, pp. 467-75.

[32]Rabindranath Tagore, *op. cit.,* p. 253.

[33]Albert Schweitzer, *Pilgrimage to Humanity, trans. Walter E. Stuermann,* (New York: Philosophical Library, 1961), p. 100.

PART 2

[1]H.G. Wells, quoted in *The New York Times Magazine,* February 20, 1955, p. 24.

[2]Ina Corrinne Brown, *Understanding Race Relations* (Englewood Cliffs, N.J.: Prentice-Hall, 1973), p. 44.

[3]Whitney M. Young, Jr., *Beyond Racism: Building an Open Society* (New York: McGraw-Hill, 1969), p. 73.

[4]Benjamin Disraeli, *Commons,* February 1, 1849.

[5]Brown, *op. cit.,* p. 8.

[6]*Ibid.,* p. 22.

[7]Stanley C. Scott, "Presidential Aide Reminds Military Much to be Done in Race Relations," *Los Angeles Sentinel,* November 8, 1973.

[8]Charles Jeffrey, "The Black Soldier: How Good is He?" Tulsa, Oklahoma, *Eagle,* October 4, 1973.

[9]Thomas Paine, quoted in George Seldes, ed., *Great Quotations* (New York: Pocket Books, 1967), p. 849.

[10]Jeffrey, *loc. cit.*

[11]Mamie Fujiyama, "Letters to the Editor," *Tulsa Daily World,* June 2, 1974.

[12]Anthony Griggs, "Minorities in the Armed Forces," *Race Relations Reporter,* July, 1973.

[13]*Ibid.*

[14]Daniel Webster, address delivered at Bunker Hill Cornerstone Laying, June 17, 1825.

[15]Paul M. Syscavage and Earl F. Mellor, "The Economic Situation of Spanish Americans," *Monthly Labor Review,* April, 1973, pp. 3-9.

[16]Representative Chet Holifield, quoted in *Washington Post,* September 13, 1973.

[17]Octavio Paz, *The Labyrinth of Solitude* (New York: Grove Press, 1961), p. 25.

[18]*Air Force Policy Letter for Commanders,* Number 12, December, 1973.

[19]*Chief of Staff Regulation 15-11,* August 20, 1973.

[20]Donald Miller, quoted in *Black Enterprise,* March, 1973, p. 58.

[21]*Ibid.,* p. 61.

[22]"Race Relations on an Aircraft Carrier," *All Hands,* November, 1973.

[23]Quoted in *Indianapolis Record,* April 21, 1973.

[24]Ed Castillo, "Employee at Kelly Files Suit," *San Antonio Light,* November 12, 1973.

[25]*Ibid.*

[26]Jeremiah O'Leary, "Bias Toward Black Officer for Post in Chile Denied," *Washington Star News,* November 30, 1973.

[27]*Washington Post,* December 1, 1973.

[28]Richard Halloran, "U.S. Army Division in Korea Combats Racial Flare-up," *The New York Times,* December 4, 1973.

[29]"Black Ex-POW Sergeant Raps His Welcome Home," *Stars and Stripes,* September 7, 1973.

[30]*Washington Star News,* August 28, 1973.

[31]"Navy Reports Racial Strife," *Overseas Weekly,* March, 1974.

[32]Griggs, *loc. cit.*

[33]Jon. M. Samuels, "Systematic Resolution of Social Problems Within the Military," *Air University Review,* July-August, 1973, pp. 73-79.

[34]Robert A. Martin, Jr., "Progress Termed Damned Serious Failure," *Overseas Weekly,* May 28, 1973.

[35]Quoted by Larry Phillips, "Race Bias Still a Problem," *Army Times,* May 23, 1973.

[36]Ron Sanders, "Race Institute Grads Eye Future," *Air Force Times,* March 28, 1973.

[37]Jeff Jones, "Navy Seminars on Race Scheduled for 170,000," *Navy Times,* March, 1973.

[38]John Young, "Marines Tackle Racism on Okinawa," *Stars and Stripes,* September, 1973.

[39]R. Drew Upright, "Air Force Plays Prejudice Game," *Miami News,* September, 1973.

[40]John Ciardi, "Literature Undefended," *Saturday Review,* January 31, 1959, p. 22.

PART 3

[1]Christabel Pankhurst, quoted in George Seldes, ed., *Great Quotations* (New York: Pocket Books, 1967), p. 982.

[2]Martin Luther, quoted in Seldes, *op. cit.,* p. 981.

[3]Sarah M. Grimké, letter addressed to Mary S. Parker, 1837, quoted in Miriam Schneir, ed., *Feminism: The Essential Historical Writings* (New York: Random House, 1972), p. 37.

[4]Elizabeth Cady Stanton, Susan B. Anthony, and Matilda Joslyn Gage, *The History of Woman Suffrage, Vol. I* (New York: National American Woman Suffrage Association, 1922), pp. 60-61.

[5]Catherine A. Beecher, *Essay on Slavery and Abolitionism, with Reference to the Duty of American Females* (Philadelphia: Henry Perkins, 1837), pp. 99.

[6]Carrie Chapman Catt, quoted in *Colliers' Year Book, 1971* (New York; Crowell Collier, 1971), p. 58.

[7]John Stuart Mill, "The Subjection of Women," in Schneir, *op. cit.,* p. 162.

[8]Sigmund Freud, quoted in Seldes, p. 87.

[9]Mill, *op. cit.,* p. 178.

[10]See Kathleen M. Snow, "My Liberated Mind Has A Wuthering Heights Heart," *Harper's,* July, 1973, p. 87.

[11]Pat Crigler, quoted by Helen Call, "Why Does Activist Feminist Role Attract or Repel Some Women?" *San Diego Union,* Oct. 21, 1973.

[12]*Ibid.*

[13]*Ibid.*

[14]*Colliers' Year Book, 1970,* p. 60.

[15]Caroline Bird. *Born Female* (New York: David McKay, 1967), p. xi.

[16]Betty Friedan, *The Feminine Mystique* (New York: W.W. Norton, 1963), p. 15.

[17]Quoted in *Ms.,* July, 1972, p. 23.

[18]Catharine Stimpson, "Thy Neighbor's Wife, Thy Neighbor's Servants: Women's Liberation and Black Civil Rights," in Vivian Gornick and Barbara K. Moran, eds., *Women in Sexist Society* (New York: Basic Books, 1971), p. 452.

[19]Shulamith Firestone, *The Dialectic of Sex: The Case for Feminist Revolution* (New York: William Morrow, 1970), pp. 16-45.

[20]Bird, *op. cit.,* pp. 262-63.

[21]*Today,* the number of women-oriented reading materials is in greater demand by publishers than materials about ethnic minorities.

[22]William Henry Chafe, *The American Woman: Her Changing Social, Economic, and Political Roles, 1920-1970,* (New York: Oxford University Press, 1972), p. 227.

[23]Simone de Beauvoir, *The Second Sex,* trans. H.M. Parchley, (New York: Alfred A. Knopf, 1952), p. 731.

[24]George Gilder, "The Suicide of Sexes," *Harper's,* June, 1973, pp. 42-54.

[25]Shulamith Firestone, "On American Feminism," in Gornick and Moran, *op. cit.,* p. 485.

[26]Kate A. Arbogast, "A Rediscovered Resource," *Military Review,* November, 1973.

[27]*Ibid.*

[28]"Marriage and Pregnancy Curbs on WAC Enlistees are Eased," *New York Times,* March 27, 1973.

[29]"The Expanding Role of Navy Women," *All Hands,* April 1, 1973.

[30]"Women Gain Sex Equality in University of Vermont R.O.T.C.," *New York Times,* June 5, 1973.

[31]"Now Military Is Putting Women into 'Men Only' Jobs," *U.S. News and World Report,* December, 1973, pp. 82-84.

PART 4

[1]Bertolt Brecht, "Mother Courage and Her Children," in Henry Hewes, ed., *The Ten Best Plays of 1962-1963* (New York: Dodd, Mead, 1963), p. 191.

[2]William B. Aycock and Seymore W. Wurfel, *Military Law Under the Uniform Code of Military Justice* (Chapel Hill: University of North Carolina Press, 1955), p. 3.

[3]John R. Thornock, "Military Trials of Civil Crimes," *Military Review,* **51** (December, 1971), 88-89.

[4]George Walton, *The Tarnished Shield: A Report on Today's Army* (New York: Dodd, Mead, 1973), pp. 195-96.

[5]Joseph W. Bishop, "The Quality of Military Justice," *New York Times Magazine,* February 22, 1970, p. 33.

[6]Thornock, *op. cit.,* p. 89.

[7]"U.S. Military Justice on Trial," *Newsweek,* August 31, 1970, p. 20.

[8]*Ex Parte Milligan,* USSC, 4 Wall., Vol. 71, 1866.

[9]*Reeves* v. *Ainsworth,* USSC, Vol. 219, 1911.

[10]*O'Callahan* v. *Parker,* USSC, Vol. 395, 1969.

[11]Harold L. Miller, "Court-Martial Jurisdiction Since O'Callahan," *Military Review,* **51** (January, 1971), 78.

[12]Saul Levitt, *The Andersonville Trial: A Play in Two Acts* (New York: Random House, 1960), p. 120.

[13]Thomas B. Macaulay, quoted in Bishop, *op. cit.,* p. 36.

[14]Dwight D. Eisenhower, quoted in *Newsweek, op. cit.,* p. 18.

[15]Senator Sam Ervin, quoted in Robert Sherrill, *Military Justice is to Justice as Military Music is to Music* (New York: Harper and Row, 1969), p. 68.

[16]Walton, *op. cit.,* p. 201.

[17]Bohdan Prehar, "In Defense of the Military Justice System," *Military Review,* **53** (January, 1973), 35.

[18]Bishop, *op. cit.,* p. 36.

[19]Prehar, *op. cit.,* p. 36.

[20]James A. Mounts and Myron G. Sugarman, "A Renaissance for Military Justice," *Military Review,* **49** (September, 1969), 9.

[21]Sherrill, *op. cit.,* p. 2.

[22]Justice Hugo Black, quoted in Bishop, *op. cit.,* p. 35.

[23]Edward F. Sherman, "Military Injustice," *New Republic,* March 3, 1968, p. 22.

[24]Sherrill, *op. cit.,* p. 63.

[25]F. Lee Bailey, quoted in *Newsweek, op. cit.,* p. 22.

[26]Bishop, *op. cit.,* p. 37.

[27]Sherrill, *op. cit.,* p. 77.

[28]*Newsweek, loc. cit.*

[29]*Ibid.*

[30]Sherman, *op. cit.,* p. 21.

[31]Joseph Heller, *Catch-22* (New York: Dell, 1961), pp. 79-80.

[32]*Newsweek, op. cit.,* p. 18.

[33]Heller, *op. cit.,* p. 180.

[34]Sherrill, *op. cit.,* p. 216.

[35]Maryann Weissman, "Court Martial," in Jonathan Black, ed., *Radical Lawyers: Their Role in the Movement and in the Courts* (New York: Avon Books, 1971), p. 160.

[36]General David M. Shoup, quoted in Sherrill, *op. cit.,* p. 213.

[37]Ossie Davis, "Purlie Victorious," in Alexander W. Allison *et. al.,* eds., *Masterpieces of the Drama,* 3rd ed. (New York: Macmillan, 1974), p. 925.

[38]"The Army: Clearing the Record," *Newsweek,* October 16, 1972, p. 36. See also, "U.S. Army Clears Black Soldiers of 66-Year-Old Crime," *Ebony,* March, 1971, pp. 31-39.

[39]John D. Weaver, *The Brownsville Raid* (New York: W. W. Norton, 1970), p. 6.

[40]*Ebony, op. cit.,* p. 37.

[41]"Final Justice to Black GI Recalls Texas Fort Saga," *Dallas Morning News,* April 30, 1973. See also "Brownsville Survivor to get $25 Thousand; Wives Maybe $10 Thousand," *Jet Magazine,* December 6, 1973.

[42]"Insist that Black Sailors Accused in Navy Riots Did Not Get Equal Justice," *Indianapolis Recorder,* April 21, 1973.

[43]Anthony Griggs, "Minorities in the Armed Forces," Part I, *Race Relations Reporter,* July, 1973, p. 9.

[44]*Ibid.,* p. 10.

[45]"U.S. Sailors in Japan Jump Over Alleged Bias and Severity," *New York Times,* June 16, 1974.

[46]"Marine Who Complained of Bias Faces Court-Martial Over His Hair," *Philadelphia Bulletin,* May 24, 1973, May 24, 1973.

[47]*Ibid.*

[48]Griggs, *loc. cit.*

[49]Davis, *op. cit.,* p. 923.

[50]Lewis C. Olive, "Duty, Honor, Country, But No Reward," *Louisville Defender,* May 10, 1973.

[51]Griggs, *op. cit.,* p. 11.

[52]"Concern Grows for Blacks Discharged in 'Other-Than-Honorable' Circumstances," *Milwaukee Star-Times,* August 9, 1973.

[53]*Ibid.*

[54]Sherrill, *op. cit.,* p. 219.

[55]Anthony Griggs, "Minorities in the Armed Forces," Part II, *Race Relations Reporter,* September, 1973, p. 9.

[56]Griggs, Part I, *op. cit.,* p. 11.

[57]Griggs, Part II, *op. cit.,* p. 11.

[58]"Military Justice Task Force Report," *Commanders Digest Summary,* March 22, 1973, p. 1.

[59]"Servicemen Petition Dellums to Abolish 'Captain's Mast'," *Oakland Tribune,* October 13, 1973.

[60]*Commanders Digest Summary, op. cit.,* p. 15.

[61]*Ibid.*

[61]Griggs, Part II, *loc. cit.*

[63]"Navy Lengthens Training to Teach Men Rules, Law," *Washington Star-News,* September 5, 1973.

[64]Griggs, Part II, *loc. cit.*

[65]Walton, *op. cit.,* p. 67.

PART 5

[1]Johann Wolfgang von Goethe, *Faust,* Part I, trans. Louis MacNeice (New York: Oxford University Press, 1969).

[2]"$25 Billion Down the Hatch," *Tulsa Tribune,* July 10, 1974.

[3]Smithsonian Institution, *Drugs in Perspective: A Fact Book on Drug Use and Misuse* (Washington, D.C.: Smithsonian Press, 1972), p. 45.

[4]"3,647 Nuclear Weapons Employees Relieved of Jobs for Drug, Alcohol, Other Problems," *Drugs and Drug Abuse Education Newsletter,* January, 1974, p. 8.

[5]Harris Isbell, "Medical Aspects of Opiate Addiction," in John A. O'Donnell and John C. Ball, eds., *Narcotic Addiction* (New York: Harper and Row, 1966), pp. 68-69.

[6]Quoted in John R. Williams, *Narcotics and Drug Dependence* (Beverly Hills: Glencoe Press, 1974), pp. 22-23.

[7]Melvin H. Weinswig, *Use and Misuse of Drugs Subject to Abuse* (New York: Pegasus, 1973), p. 23.

[8]Morris M. Rubin, "Panel Discussion on Drug Addiction," in James C. Bennett and George D. Demos, eds., *Drug Abuse and What We Can Do About It* (Springfield, Ill.: Charles C. Thomas, 1970), p. 27.

[9]Williams, *op. cit.,* p. 22.

[10]Rubin, *op. cit.,* p. 27.

[11]Alfred R. Lindesmith, *Addiction and Opiates* (Chicago: Aldine Press, 1968), p. 64.

[12]Johannes Biberfield, quoted in Lindesmith, *ibid.,* p. 65.

[13]Isbell, *op. cit.,* p. 70.

[14]Alfred R. Lindesmith, "Basic Problems in the Social Psychology of Addiction and a Theory," in O'Donnell, *op. cit.,* p. 99.

[15]Jerome H. Jaffe, Pharmacological Approaches to the Treatment of Compulsive Opiate Use: Their Rationale and Current Status," in Perry Black, ed., *Drugs and the Brain: Papers on the Use and Abuse of Psychotropic Agents* (Baltimore: John Hopkins University Press, 1969), p. 352.

[16]Lindesmith, *Addiction and Opiates, op. cit.,* pp. 95-96.

[17]Thomas De Quincey, *Confessions of an English Opium-Eater* (Edinburgh: Adam and Charles Black, 1862), p. 275.

[18]Lindesmith, "Basic Problems. . .," *op. cit.,* p. 103.

[19]Howard S. Becker, "Becoming a Marijuana User," in O'Donnell, *op. cit.,* p. 121.

[20]Jack Gebler, *The Connection* (New York: Grove Press, 1957), pp. 31-32.

[21]Dorothy Nelkin, *Methadone Maintenance: A Technological Fix* (New York: George Braziller, 1973), p. 18.

[22]Bernard Langenaver, "A Follow-up Study of Narcotics Addicts in the NARA Program," *American Journal of Psychiatry,* **128** (July, 1971), 74-75.

[23]Dan Waldorf, *Careers in Dope* (Englewood Cliffs, N.J.: Prentice-Hall, 1973), p. 31.

[24]M.A. Farber, "Veterans Still Fight Vietnam Drug Habits," *New York Times,* June 2, 1974.

[25]Norman E. Zinberg, "Rehabilitation of Heroin Users in Vietnam," *Contemporary Drug Problems, 1* (Spring, 1972), 263.

[26]Marie Nyswander, quoted in Nat Hentoff, *A Doctor Among the Addicts* (New York: Grove Press, 1970), pp. 76-77.

[27]Waldorf, *op. cit.,* p. 37.

[28]Farber, *op. cit.*, p. 46.

[29]Waldorf, *loc. cit.*

[30]*Ibid.*

[31]Nelkin, *op. cit.*, p. 122.

[32]Charles E. Goshen, *Drinks, Drugs, and Do-Gooders* (New York: The Free Press, 1973), p. 150.

[33]Maryanne V. Looney and Suzanne Metcalf, quoted in "External Pressures More Important Than Internal Ones in Motivating Addicts to Seek Treatment, Researchers Contend," *Drugs and Drug Abuse Education Newsletter,* March, 1974, p. 5.

[34]Father Patrick O'Connor, quoted in Rubin, *op. cit.*, p. 46.

[35]Thomas F.A. Plaut, *Alcohol Problems: A Report to the Nation by the Cooperative Commission on the Study of Alcoholism* (New York: Oxford University Press, 1967), p. 39.

[36]Jay N. Cross, *A Guide to the Community Control of Alcoholism* (Washington, D.C.: American Public Health Association, 1968), p. 50.

[37]Sydney Cahn, *The Treatment of Alcoholism: An Evaluative Study* (New York: Oxford University Press, 1970), pp. 36-37.

[38]Eva Maria Blum and Richard H. Blum, *Alcoholism: Modern Psychological Approaches to Treatment* (San Francisco: Jossey-Bass, 1967), p. 44.

[39]Joseph B. Kendis, "The Human Body and Alcohol, " in David J. Pittman, ed., *Alcoholism* (New York: Harper & Row, 1967), pp. 28-29.

[40]Henry Wechsler *et. al.*, "Religious-Ethnic Differences in Alcohol Consumption," *Journal of Health and Social Behavior,* **2** (March, 1970), 28-29.

[41]David J. Pittman, "International Overview: Social and Cultural Factors in Drinking Patterns, Pathological and Nonpathological," in Pittman, *op. cit.*, p. 9.

[42]Ronald J. Catanzaro, "Psychiatric Aspects of Alcoholism," in Pittman, *op. cit.*, pp. 36-37.

[43]Louis A. Faillace and Robert F. Ward, "Psychiatric Approaches to Alcoholism," in Black, *op. cit.*, p. 333.

[44]David A. Overton, "State-Dependent Learning Produced by Alcohol," in Benjamin Kissen and Henri Begleiter, eds., *The Biology of Alcoholism: Vol. 2: Physiology and Behavior* (Baltimore: John Hopkins University Press, 1969), p. 333.

[45]Catanzaro, *op. cit.*, p. 38.

[46]Goshen, *op. cit.,* pp. 99.

[47]John L. Horn, *et al.,* "Diagnosis of Alcoholism: Factors of Drinking, Background, and Current Conditions of Alcoholics," *Studies in Alcohol,* **35** (March, 1974), 172.

[48]James C. Bennett, "What is Normal?," in Bennett and Demos, *op cit.,* p. 91.

[49]Thomas Edward Bratter, "Treating Alienated, Unmotivated, Drug Using Adolescents," *American Journal of Psychotherapy,* **27** (October, 1973), 586-87.

[50]Zinberg, *op. cit.,* p. 769.

[51]"Drug Abuse Survey Made in Six New Jersey High Schools," *Health Services Reports,* **67** (May, 1972), 416-17.

[52]Edward A. Suchman, "The 'Hang-loose' Ethic and the Spirit of Drug Use," *Journal of Health and Social Behavior,* **9** (June, 1968), 148-55.

[53]Paul M. Kohn and G.W. Mercer, "Drug Use, Drug Use Attitudes, and the Authoritarianism-Rebellion Dimension," *Journal of Health and Social Behavior,* **12** (June, 1971), 130.

[54]Gabriel G. Nohans, *Marihuana: Deceptive Weed* (New York: Raven Press, 1973), p. 263.

[55]Troy Winslow, *et. al.,* "Drug Involvement: A Response to Inadequate Environment," *Journal of Drug Education,* **1** (September, 1972), 276-77.

[56]Marlin H. Dearden, "Adolescent Drug Abuses — A Problem of Interpersonal Relations and School Organization," *Journal of Drug Education,* **1** (September, 1971), 210.

[57]Nohans, *op. cit.,* p. 260.

[58]Richard Brotman, "Drug Abuse: The Dilemma of Criminal-Sick Hypothesis," in Black, *op. cit.,* p. 374.

[59]Zinberg, *op. cit.,* p. 763.

[60]Dearden, *op. cit.,* p. 214.

PART 6

[1]Quoted in Selig Greenberg, *The Quality of Mercy* (New York: Atheneum, 1971), p. 117.

[2]Harry Nelson, "No Quick Pill for Health Care," editorial in *Tulsa Daily World,* June 22, 1974.

[3]Edward Kennedy, *In Critical Condition: The Crisis in American Health* (New York: Simon and Schuster, 1972), p. 15.

[4]Ralph Nader Study, *One Life — One Physician* (Washington, D.C.: Public Affairs Press, 1971), p. 4.

[5]Richard D. Lyons, "The Coming Crisis," *Colliers Yearbook,* 1973 (New York: Macmillan Educational Corp., 1973), p. 76.

[6]Michael DeBakey, quoted in "Russ Citizens Get Better Medical Care," *Tulsa Daily World,* June 20, 1974.

[7]Bonnie Bullough and Verne L. Bullough, *Poverty, Ethnic Identity and Health Care* (New York: Meredith Corp., 1972), p. 75.

[8]*Ibid.,* p. 77.

[9]*Ibid.,* p. 106.

[10]Lyons, *op. cit.,* p. 76.

[11]Greenberg, *op. cit.,* p. 97.

[12]Rollo May, *Power and Innocence* (New York: W.W. Norton, 1972), p. 37.

[13]Lyons, *loc. cit.*

[14]"Hospitals Reported Using Unlicensed Doctors," *Tulsa Daily World,* June 21, 1974.

[15]National Commission on Community Health Services, *Health Is a Community Affair* (Cambridge: Harvard University Press, 1966), p. 2.

[16]Bernard J. Stern, *Medical Service by Government: Local, State, and Federal* (New York: Commonwealth Fund), 1946, p. 170.

[17]George W. Bachman and Associates, *Health Resources in the United States* (Washington, D.C.: Brookings Institution, 1952), p. 222.

[18]Samuel W. Bloom, *The Doctor and His Patient: A Sociological Interpretation* (New York: Russell Sage Foundation, 1963), p. 255.

[19]Hans O. Mausch, "Nursing: Churning for Change," in Howard Freeman et. al., eds., *Handbook of Medical Sociology* (Englewood Cliffs, N.J.: Prentice-Hall, 1972), p. 211.

[20]"The Nature of Nursing," *American Journal of Nursing,* **63** (August, 1964), 62-28.

[21]John Stuart Mill, quoted in John H. Knowles, *Hospitals, Doctors, and the Public Interest* (Cambridge: Harvard University Press, 1965), p. 125.

[22]Stephen P. Strickland, *U.S. Health Care: What's Wrong and What's Right* (New York: Universe Books, 1972), p. 44.

[23]Fred J. Cook, *The Plot Against the Patient* (Englewood Cliffs: N.J., Prentice-Hall, 1967), p. 26.

[24]Raymond S. Duff and August B. Hollingshead, *Sickness and Society* (New York: Harper & Row, 1968), pp. 124-25.

[25]Osofsky, *op. cit.,* p. 241.

[26]Ozzie G. Simmons, "Implications of Social Class for Public Health," *Human Organization,* **16** (1958), 16-18.

[27]Osofsky, *op. cit.,* p. 256.

[28]Sydney H. Croog and Donald F. Ver Steeg, "The Hospital as a Social System," in Freeman, *op. cit.,* p. 274.

[29]Knowles, *op. cit.,* p. 171.

[30]Croog and Ver Steeg, *op. cit.,* pp. 289-90.

[31]Senator Proxmire, quoted in "V.A. Hospital Gear Idle," *Tulsa World,* July 7, 1974.

[32]Commander Gary D. Despiegler, quoted in "Navy Hospital Has Race Rap Sessions," *Norfolk Journal and Guide,* August 18, 1973.

PART 7

[1]William James, quoted in George Seldes, ed., *The Great Quotations* (New York: Pocket Books, 1967), p. 266.

[2]John W. Macy, Jr., *Public Service: The Human Side of Government* (New York: Harper & Row, 1971), p. 3.

[3]Frederick C. Mosher, *Democracy and the Public Service* (New York: Oxford University Press, 1968), p. 57.

[4]*Ibid.,* p. 62.

[5]*Ibid.,* p. 63.

[6]Thomas Hobbes, quoted in Seldes, *op. cit.,* p. 750.

[7]Don Hellriegel and Larry Short, "Equal Employment Opportunity in the Federal Government: A Comparative Analysis," *Public Administration Review,* **32** (November-December, 1972), 851.

[8]Mosher, p. 65.

[9]*Ibid.,* p. 73.

[10]Gerald R. Ford, Address at the 91st Anniversary of the Federal Civil Service, Washington, D.C., January 16, 1974.

[11]Bernard Rosen, "Commentary," *Civil Service Journal,* **9** (January-March, 1974), 18.

[12]Honoré de Balzac, quoted in Seldes, *op. cit.,* p. 102.

[13]Mary McCarthy, *New Yorker,* October 18, 1958, p. 202.

[14]"Bearding Uncle Sam," *Time,* August 31, 1970, p. 64.

[15]"Union Membership Among Government Employees," *Monthly Labor Review,* **93** (July, 1970), 20.

[16]Louise Durbin, "Making a Federal Case Out of a B.A.," *Mademoiselle,* June, 1968, p. 138.

[17]*Time, op cit.,* p. 65.

[18]Macy, *op. cit.,* p. 4.

[19]*Ibid.,* p. 2.

[20]Daniel Seligman, "How Equal Opportunity Turned Into Employment Quotas," *Fortune,* March, 1973, pp. 160-68.

[21]"The Merit System and Equal Opportunity Employment," *Civil Rights Journal,* January-March, 1973.

[22]Richard Hebert, "Federal Job Ghetto," *Nation,* February 21, 1972, p. 242.

[23]*Ibid.,* p. 243.

[24]*Ibid.,* p. 244.

[25]Samuel Krislov, *The Negro in Federal Employment: The Quest for Equal Opportunity* (Minneapolis: University of Minnesota Press), 1967, p. 13.

[26]Earl J. Reeves, "Making Equality of Employment a Reality in the Federal Service," *Public Administration Review,* **30** (January-February, 1970), 45.

[27]Macy, *op. cit.,* pp. 74-75.

[28]Ollie A. Jensen, "Cultural Bias in Selection," in Robert T. Golembiewski and Michael Cohen, eds., *People in Public Service: A Reader in Public Personnel Administration* (Itasca, Ill.: F.E. Peacock, 1970), p. 288.

[29]*Ibid.,* p. 295.

[30]Macy, *op. cit.,* p. 79.

[31]Julius W. Hobson, "Uncle Sam is a Bigot," *Saturday Evening Post,* April 20, 1968, pp. 16-18.

[32]"Uncle Sam's Own Black Bias," *Business Week,* August 28, 1971, p. 36.

[33]Mike Causey, "Minority Employment Up by 10,000," *Washington Post,* March 5, 1974.

[34]"Gain in Minority Groups Employment," *Civil Service News,* February-March, 1974, p. 1.

[35]"Double Standard," *New Republic,* May 29, 1971, p. 12.

Moore, Joan. *Mexican-Americans.* Englewood Cliffs, N.J.: Prentice-Hall, 1970.

Office of the Assistant Secretary of Defense. *Integration of the Negro in the Armed Forces of the U.S.A.* Washington, D.C.: U.S. Government Printing Office, 1962.

Rand, Christopher. *The Puerto Ricans.* New York: Oxford University Press, 1968.

Samora, Julian. *La Raza: Forgotten Americans,* South Bend, Ind.: University of Notre Dame Press, 1966.

Steiner, Stan. *The New Indians.* New York: Dell, 1968.

Steinfield, Melvin. *Cracks in the Melting Pot: Racism and Discrimination in American History.* Beverly Hills, Ca.: Glencoe Press, 1970.

Stillman, Richard J., II. *Integration of the Negro in the U.S. Armed Forces.* New York: Frederick A. Praeger, 1968.

Theus, Lusius. *Education Program in Race Relations.* Washington, D.C.: U.S. Government Printing Office, 1970.

United States Commission on Civil Rights. *Racism in America and How to Combat It.* Washington, D.C.: U.S. Government Printing Office, 1970.

——— *Family Housing and the Negro Serviceman.* Washington, D.C.: U.S. Government Printing Office, 1963.

Young, Whitney M., Jr. *Beyond Racism: Building an Open Society.* New York: McGraw-Hill, 1969.

WOMEN'S EQUALITY

Adelstein, Michael E., and Jean G. Pival, eds. *Women's Liberation.* New York: St. Martin's Press, 1972.

Amundsen, Kirsten. *The Silenced Majority: Women and American Democracy.* Englewood Cliffs, N.J.: Prentice-Hall, 1971.

Beard, Mary R. *Women as Force in History.* New York: Collier Books, 1971.

Bird, Caroline. *Born Female: The High Cost of Keeping Women Down.* New York: David McKay, 1970.

Cross, Barbara M., ed. *The Educated Woman in America.* New York: Teachers College Press, 1965.

de Beauvoir, Simone. *The Second Sex.* H.M. Parchley, trans. New York: Alfred A. Knopf, Inc., 1952.

Decter, Midge. *The New Chastity and Other Arguments Against Women's Liberation.* New York: Coward, McCann & Geoghegan, 1972.

Epstein, Cynthia F. *Woman's Place.* Berkeley: University of California Press, 1970.

Firestone, Shulamith. *The Dialectic of Sex: The Case for Feminist Revolution.* New York: William Morrow, 1970.

Friedan, Betty. *The Feminine Mystique.* New York: W.W. Norton, 1963.

Gornick, Vivian, and Barbara K. Moran, eds. *Women in Sexist Society: Studies in Power and Powerlessness.* New York: Basic Books, 1971.

Henderson, George. *To Live in Freedom: Human Relations Today and Tomorrow.* Norman: University of Oklahoma Press, 1972.

Howard Jane. *A Different Woman.* New York: E.P. Dutton, 1973.

Koedt, Anne, Ellen Levine, and Anita Rapone, eds. *Radical Feminism.* New York: Quadrangle, 1973.

Laffin, John. *Women in Battle.* New York: Abelard Schuman, 1967.

Marine, Gene. *A Male Guide to Women's Liberation.* New York: Holt, Rinehart & Winston, 1972.

Mill, John Stuart. *On the Subjection of Women.* Greenwich, Conn.: Fawcett Publications, 1971.

Mitchell, Juliet. *Woman's Estate.* New York: Pantheon Books, 1971.

Moran, Robin, ed. *Sisterhood is Powerful: An Anthology of Writings from the Women's Liberation Movement.* New York: Random House, 1970.

Reeves, Nancy, ed. *Womankind: Beyond the Stereotypes.* Chicago: Aldine, 1971.

Reische, Diana L., ed. *Women and Society.* New York: H.H. Wilson, 1972.

Safilios-Rothschild, Constantina. *Women and Social Policy.* Englewood Cliffs, N.J.: Prentice-Hall, 1974.

Schneir, Miriam, ed. *Feminism: The Essential Historical Writings.* New York: Random House, 1972.

Sochen, June. *Movers and Shakers: Women Thinkers and Activists, 1900-1970.* New York: Quadrangle, 1973.

Stambler, Sookie, ed. *Women's Liberation: Blueprint for the Future.* New York: Ace Publishing Corp., 1970.

Stanton, Elizabeth Cady, Susan B. Anthony, and Matilda Joslyn Gage. *The History of Woman Suffrage,* Vol. I. New York: National American Woman Suffrage Association, 1922.

Tanner, Leslie B., ed. *Voices From Women's Liberation.* New York: Signet, 1970.

Theodore, Athena, ed. *The Professional Woman.* Cambridge, Mass.: Schenkman, 1971.

Thompson, Clara M. *On Women.* New York: New American Library, 1964.

MILITARY JUSTICE

Aycock, William B., and Seymore W. Wurfel. *Military Law Under the Uniform Code of Military Justice.* Chapel Hill: University of North Carolina Press, 1955.

Black, Jonathan, ed. *Radical Lawyers: Their Role in the Movement and in the Courts.* New York: Avon, 1971.

Davis, George B. *A Treatise on the Military Law of the United States.* New York: Wiley, 1909.

Finn, James. *Conscience and Command: Justice and Discipline in the Military.* New York: Random House, 1971.

Lane, Ann J. *The Brownsville Affair.* New York: Kennikat Press, 1971.

Levitt, Saul. *The Andersonville Trial: A Play in Two Acts.* New York: Random House, 1960.

McComsey, John A., and Morris O. Edwards. *The Soldier and the Law.* Harrisburg, Pa.: Military Science Publishing, 1941.

National Association for the Advancement of Colored People. *The Search for Military Justice: Report of an NAACP Inquiry into the Problems of Negro Servicemen in West Germany.* New York: NAACP Contribution Fund, 1971.

Sherrill, Robert. *Military Justice is to Justice as Military Music is to Music.* New York: Harper & Row, 1969.

Walton, George. *The Tarnished Shield: A Report on Today's Army.* New York: Dodd, Mead, 1973.

THE USE AND ABUSE OF ALCOHOL AND DRUGS

Black, Perry, ed. *Drugs and the Brain: Papers on the Use and Abuse of Psychotropic Agents.* Baltimore: Johns Hopkins University Press, 1969.

Blum, Eva Maria, and Richard H. Blum. *Alcoholism: Modern Psychological Approaches to Treatment.* San Francisco: Jossey-Bass, 1967.

Brecher, Edward M., and the Editors of Consumer Reports. *Licit and Illicit Drugs.* Boston: Little, Brown, 1972.

Cross, Jay N. *A Guide to the Community Control of Alcoholism.* Washington, D.C.: American Public Health Association, 1968.

De Quincey, Thomas. *Confessions of an English Opium-Eater.* Edinburgh, Scotland: Adam and Charles Black, 1862.

Fort, Joel. *Alcohol: Our Biggest Drug Problems.* New York: McGraw-Hill, 1973.

Gebler, Jack. *The Connection.* New York: Grove Press, 1957.

Goshen, Charles E. *Drinks, Drugs, and Do-Gooders.* New York: The Free Press, 1973.

Hardy, Richard E., and John G. Cull. *Climbing Ghetto Walls: Disadvantaged, Delinquency, and Rehabilitation.* Springfield, Ill.: Charles C. Thomas, 1973.

Hentoff, Nat. *A Doctor Among the Addicts.* New York: Grove Press, 1970.

Johnson, Bruce D. *Marihuana Users and Drug Subculture.* New York: Wiley, 1973.

Johnson, Vernon E. *I'll Quit Tomorrow.* New York: Harper & Row, 1973.

Kissen, Benjamin, and Henri Begleiter, eds. *The Biology of Alcoholism,* Vol. 2. Baltimore: Johns Hopkins University Press, 1969.

Lindesmith, Alfred R. *Addiction and Opiates.* Chicago: Aldine, 1968.

Luft, Joseph. *Group Processes: An Introduction to Group Dynamics.* 2d Edition. Palo Alto, Ca.: National Press Books, 1970.

Nelkin, Dorothy. *Methadone Maintenance: A Technological Fix.* New York: George Braziller, 1973.

Nohans, Gabriel G. *Marihuana: Deceptive Weed.* New York: Raven Press, 1973.

O'Donnel, John A., and John C. Ball, eds. *Narcotics Addiction.* New York: Harper & Row, 1966.

Ohlsen, Merle M. *Group Counseling.* New York: Holt, Rinehart & Winston, 1970.

Pittman, David J., ed. *Alcoholism.* New York: Harper & Row, 1967.

Plaut, Thomas F.A. *Alcohol Problems: A Report to the Nation by the Cooperative Commission on the Study of Alcoholism.* New York: Oxford Universtiy Press, 1967.

Seymour, Whitney N. *The Young Die Quietly: The Narcotics Problem in America.* New York: Morrow, 1971.

Siegel, Harvey H. *Alcohol Detoxification Programs: Treatment Instead of Jail.* Springfield Ill.: Charles C. Thomas, 1973.

Steiner, Claude. *Games Alcoholics Play: The Analysis of Life Scripts.* New York: Grove, 1971.

Smithsonian Institution. *Drugs in Perspective: A Fact Book on Drug Use and Misuse.* Washington, D.C.: Smithsonian Press, 1972.

United States Commission on Marihuana and Drug Abuse. *Marihuana: A Signal of Misunderstanding; First Report.* Washington, D.C.: U.S. Government Printing Office, 1972.

Waldorf, Dan. *Careers in Dope.* Englewood Cliffs, N.J.: Prentice-Hall, 1973.

Weinswig, Melvin H. *Use and Misuse of Drugs Subject to Abuse.* New York: Pegasus, 1973.

Williams, John R. *Narcotics and Drug Dependence.* Beverly Hills, Ca.: Glencoe Press, 1974

HEALTH CARE

Bachman, George W., and Associates. *Health Resources in the United States.* Washington, D.C: Brookings Institution, 1952.

Bloom, Samuel W. *The Doctor and His Patient: A Sociological Interpretation.* New York: Russell Sage Foundation, 1963.

Brown, Esther L. *Newer Dimensions of Patient Care.* New York: Russell Sage Foundation, 1965.

Bullough, Bonnie, and Verne L. Bullough. *Poverty, Ethnic Identity and Health Care.* New York: Meredith Corp., 1972.

Cantril, Albert H., and Charles W. Roll, Jr. *Hopes and Fears of the American People.* New York: Universe Books, 1972.

Chase, Stuart. *Power of Words.* New York: Harcourt, Brace, 1954.

Coe, Rodney M., ed. *Planned Change in the Hospital.* Englewood Cliffs, N.J.: Prentice-Hall, 1970.

Cook, Fred J. *The Plot Against the Patient.* Englewood Cliffs, N.J.: Prentice-Hall, 1967.

Davitz, Lois J. *Interpersonal Processes in Nursing: Case Histories.* New York: Springer, 1970.

Dugg, Raymond S., and August B. Hollingshead. *Sickness and Society.* New York: Harper & Row, 1968.

Fast, Julius. *Body Language.* New York: M. Evans, 1970.

Freeman, Howard, Sol Levine, and Leo G. Reeder, eds. *Handbook of Medical Sociology.* Englewood Cliffs, N.J.: Prentice-Hall, 1972.

Garrett, Annette. *Interviewing: Its Principles and Methods.* New York: Family Service Association of America, 1947.

Gilbert, Ruth. *The Public Health Nurse and Her Patient,* rev. ed. Boston: Harvard University Press, 1950.

Glasser, William A. *Social Settings and Medical Organization.* New York: Atherion Press, 1970.

Greenberg, Selig. *The Quality of Mercy.* New York: Atheneum, 1971.

Gregg, Elinor D. *The Indians and the Nurse.* Norman: University of Oklahoma Press, 1965.

Hall, Edward R. *The Silent Language.* Greenwich, Conn.: Fawcett, 1959.

Hanlon, John J. *Principles of Public Health Administration.* St. Louis: C.V. Mosby, 1964.

Hayakawa, S.I. *Language in Thought and Action.* New York: Harcourt, Brace, 1964.

Huxley, Aldous. *Words and Their Meaning.* Los Angeles: Jake Zeitlin, 1940.

Kennedy, Edward. *In Critical Condition: The Crisis in American Health.* New York: Simon and Schuster, 1972.

Knowles, John H. *Hospitals, Doctors, and the Public Interest.* Cambridge, Mass.: Harvard University Press, 1965.

Koos, Earl L. *The Health of Regionville.* New York: Columbia University Press, 1954.

National Commission on Community Health Services. *Health Is a Community Affair.* Cambridge, Mass.: Harvard University Press, 1966.

Nichols, Ralph G., and Leonard A. Stevens. *Are You Listening?* New York: McGraw-Hill, 1949.

Osofsky, Howard J., ed. *Among the People: Encounters With the Poor.* New York: Basic Books, 1968.

Ralph Nader Study. *One Life — One Physician.* Washington, D.C.: Public Affairs Press, 1971.

Reik, Theodore. *Listening With the Third Ear.* New York: Farrar, Strauss, 1949.

Robinson, Lisa. *Psychological Aspects of the Care of Hospitalized Patients.* F.A. Davis, 1968.

Schorr, Daniel. *Don't Get Sick in America.* Nashville: Aurora Publishers, 1970.

Stern, Bernhard J. *Medical Services by Government: Local, State, and Federal.* New York: Commonwealth Fund, 1946.

Strickland, Stephen P. *U.S. Health Care: What's Wrong and What's Right.* New York: Universe Books, 1972.

Windemuth, Audrey. *The Nurse and the Outpatient Department.* New York: Macmillan, 1957.

CIVIL SERVICE

Cahn, Frances. *Federal Employees in War and Peace: Selection, Placement, and Removal.* Washington, D.C.: Brookings Institution, 1949.

Ekirch, Arthur A. *The Civilian and the Military.* New York: Oxford University Press, 1956.

Golembiewski, Robert T., and Michael Cohen, eds. *People in Public Service: A Reader in Public Personnel Administration.* Itasca, Ill.: F.E. Peacock, 1970.

Hickman, Martin B. *The Military and American Society.* Beverly Hills, Ca.: Glencoe Press, 1971.

Hollander, Herbert S. *Crisis in the Civil Service.* Washington, D.C.: Current Issues Publishers, 1955.

Huntington, Samuel P. *The Soldier and the State: The Theory and Politics of Civil-Military Relations.* Cambridge, Mass.: Harvard University Press, 1957.

Kammerer, Gladys M. *Impact of War on Federal Personnel Administration, 1939-1945.* Lexington: University of Kentucky Press, 1951.

Kilpatrick, Franklin P., and M. Kent Jennings. *The Image of Federal Service.* Washington, D.C.: Brookings Institution, 1964.

——— *Source Book of a Study of Occupational Values and the Image of the Federal Service.* Washington, D.C.: Brookings Institution, 1964.

Krislov, Samuel. *The Negro in Federal Employment: The Quest for Equal Opportunity.* Minneapolis: University of Minnesota Press, 1967.

Larson, Arthur D., ed. *Civil-Military Relations and Militarism: A Classified Bibliography Covering the United States and Other Nations of the World.* Manhattan: Kansas State University Library, 1971.

Macy, John W. *Public Service: The Human Side of Government.* New York: Harper & Row Publishers, 1971.

Mainzer, Lewis C. *Political Bureaucracy.* Glenview, Ill.: Scott, Foresman, 1973.

Mosher, Frederick C. *Democracy and the Public Service.* New York: Oxford University Press, 1968.

Passett, Barry A. *Leadership Development for Public Service.* Houston: Gulf Publication, 1971.

Rosenbloom, David H. *Federal Service and the Constitution.* Ithaca, N.Y.: Cornell University Press, 1971.

Smith, Darrell H. *The United States Civil Service Commission: Its History, Activities, and Organization.* Baltimore: Johns Hopkins Press, 1928.

Smith, Louis. *American Democracy and Military Power: A Study of Civil Control of the Military Power in the United States.* Chicago: University of Chicago Press, 1951.

Stahl, Oscar G. Public Personnel Administration. New York: Harper & Row, 1971.

White, Leonard D. *Civil Service in Wartime.* Chicago: University of Chicago Press, 1945.

Index